Randolph's Shop

by J. Randolph Bulgin

Bulgin Forge Press
Franklin, N.C.

ISBN 0-9785475-0-0

Printed in the United States of America
by Quebecor World Book Services

Introduction by Neil Knopf
Design by Barbara McRae

"UNTITLED" 1985   3/10   JAMES M. STOELKER

John Alexander Bulgin
Dec. 9, 1912 – July 3, 1995
My Father

Any job undertaken is a reflection of the soul of the undertaker. The skill and attention given to the aspects of the work, even those parts that are hidden from view, is an everlasting reflection of the hands and the mind that brought the work to completion.

Wood, steel, plexi, oil

in rooms large and small;

A place where hands

Work, learn, craft and toil;

A domain known to craftsmen all.

In places new

and places old,

those of craftsmen's view

find a common home,

Ozone smell from welder's arc,

coppery taste of metal flake;

skilled, learning, teaching hands form

more than mere projects;

Legacies are built,

everlasting are born.

By my son, George A. Bulgin

ACKNOWLEDGMENTS

"Given enough talented and generous friends, there is nothing that a person cannot accomplish." I said that and I believe it. And this book proves it. It is always dangerous, at least for me, to try to thank publically any group of participants in an endeavor such as this book has been for me because I will inevitably leave out some person or persons who I will wish later had been included. But I think I will risk it here because without the encouragement and help these people provided this thing would not exist.

Fred Alexander, who spent many Saturday mornings at my shop taking photographs. When I got to the point where I needed a photographer during the week while he was at his day job, he helped me to select a camera and then taught me some of the basics of using it. And then when I learned all I needed to know and subsequently fired him we still remained friends. That's a pal!

Barbara McRae, editor of our local newspaper, who took a bunch of photographs, drawings, raw text and data and transformed it into a coherent piece of work which made me look as if I know more about what I am doing than I really do.

Fred Bulgin, my younger brother, who helped me with several of the projects by doing some of the time consuming welding and also helping with the photography.

Bill Bowser, whose expertise in the use of AutoCAD proved to be indispensable to this project.

And my wife, Nan. How she puts up with me I will never know. Thanks, Nan.

INTRODUCTION

From the dawn of the industrial age, inventors, engineers, and machinists have documented their crafts through thousands of writings covering their processes, designs and ideas. Many volumes have been written covering virtually all conceivable forms of mechanical devices, artistic metalworking endeavor, and process. The readiness of those authors to put forth their knowledge in print is inspirational. Strong convictions are a must and a large part of the equation lies in the authors' willingness to share knowledge with their fellow craftsmen.

The design of mechanical apparatus requires sound knowledge of the medium used, a working knowledge of mechanical devices and a certain amount of creativity. Most people acquire one or two of these traits occasionally, all three are acquired in one person. Randolph has acquired all three. Being the fourth generation of a family ensconced in metalwork and engineering has played a large part in the formation of the author's life. After serving in the United States Navy, Randolph went on to become a machinist, welder, toolmaker, shop teacher, blacksmith, businessman and author. A half dozen of his articles have already graced the pages of *The Home Shop Machinist* and *Machinist's Workshop* magazines.

Combining his fertile mind with capable hands, Randolph has managed to build and assemble one of the finest home shops in the nation, a shop that has been source of both income and pleasure. Not being satisfied with the compromises often associated with the acquisition of off-the-shelf machinery, numerous pieces of equipment were built by the author in his shop. The devices were fabricated with function, performance and longevity being the design criteria. Randolph's shop is one of the most organized shops that can be found and is a marvel in its own right.

One does not need to read far into this work to realize that Randolph's approach to the many facets of metalworking is easily understood. The author has cunningly entwined his sensible approach, easy manner and sense of humor into an informative and entertaining presentation aimed to please the novice machinist as well as the master craftsman. The years of shop experience are readily recognized by the approach taken to solve the tasks at hand. Randolph takes the time to not only show the project but to explain the different aspects of each step. The result is an understandable, thought process-building exercise that allows the reader to expand their machining capability. Randolph's style of writing inspires the reader to believe in their own skills, to build upon their knowledge and to open new doors into the world of metal shop work.

Neil Knopf, Editor

The Home Shop Machinist / Machinists Workshop

CONTENTS

PREFACE

The skills learned and used by my grandfather as a blacksmith in the early years of the twentieth century are now practiced only as a novelty and only by people whose hobby is the pursuit and preservation of those skills. Those early blacksmith shops are gone now and have given way to modern ways and that is the way of the world.

The small, independent machine and fabrication shop where "conventional" tools and methods are employed is going down the same path as the blacksmith shop. It is an inevitable evolution and although some of us will mourn the passing of these shops from our mainstream society they will eventually exist only in the basements and backyards of the hobbyist. And therein lies the strength of the metal working trades.

The metal working hobbyist — the dedicated hobbyist in any field really, but here we are concerned with the metal working trades — is a force that will be with us for generations. I have visited other people's shops and have invited other people to visit me in my shop and the diversity, the complexity and the abilities demonstrated in some of these places and by some of these people challenges the imagination.

My father started teaching my brother and me how to weld before we were 10 years old, just as his father started teaching him and his brother blacksmithing when they were young. I was a machinist in the Navy in the early 1960s, not a Machinist's Mate but a Machinery Repairman for those of you who are also ex-Navy. I have been certified as a weldor on nuclear power systems and I have paid my dues by getting down into the ditch and welding the bottom side of the pipe. I am a member of a prestigious organization of craftsmen in the field of artist blacksmiths. I have seen CNC come of age and have learned, slowly and sometimes reluctantly, to adapt to the changes we have all seen in the metal working processes in this country within the past 50 years or so.

I list some of the things here that I have learned and know about because to list the inverse, those things that I have yet to learn, would require far more time and space than is available in this book. Every project that I undertake relating to metal working demonstrates to me more things that I do not know and, hopefully, provides me with incentive to learn them.

The day that I feel that I have no more to learn will be the day when I go to occupy the space out at Woodlawn Cemetery which, by the way, I have already paid for.

The places and opportunities available to metal workers now for learning are so much more numerous and of so much better quality than was available when I was coming along that it is mind-boggling. There are clubs, hobby groups and professional associations. There are fine publications for the hobbyist like *Home Shop Machinist, Machinist's Workshop* and *Live Steam*. There are many professional journals available including but certainly not limited to, *American Machinist, Modern Machine Shop* and *Practical Welding Today*. There are internet discussion forums where novices and experts alike can share knowledge and experiences. And these listed here are but a few of the literally thousands of sources for learning and exchanging information.

I once read a statement, I think on one of the internet bulletin board forums, by a machinist who claimed that he knew his stuff and that he was as smart as Hell. I can't dispute that because I don't know the man and I don't know just how smart that is. But I do know this: I know that in order to ask a question you must first admit that there is something you do not know. And I have found over the years that I can learn more from the man who will admit to his ignorance and help you to learn than from the man who already has all of the answers.

It is not my intent here to malign genuine expertise. The truly great mentors are the folks who are willing to share their knowledge and experience, but who know that they can also learn from the student. It is when the teacher assumes that he has no more to learn that he loses his effectiveness as a teacher. He also loses his credibility as far as I am concerned.

So be advised. There are going to be some things in this book that may be useful to some of you, but I doubt there will be any revelations for all of you. And I ask you this. Teach me what you know so that we may both learn.

R.B.

Chapter 1

The Shop

The shop. For me those have always been magic words. My earliest memories are of the reflections of a welding arc in the leaves of a maple tree which stood near the house where I grew up. The welding machine, a motor generator set powered by a six cylinder Chrysler engine, was parked underneath a shed beside my father's shop and the exhaust pipe came out through a neatly cut round hole in the back of the shed. When my father nodded his welding hood down in preparation for striking the arc, he cautioned me not to look and when I looked away as he welded, I could see the reflections of the arc in the leaves of that tree. I can see them as clearly in my mind's eye today as I could then as a 2- or 3-year-old boy. But I have seen a lot of welding arcs since that time.

My father and his brother were the owners of their shop, Bulgin Brothers Machine Shop, in the late 1930s when the threat of World War II was beginning to be in the news. They had taken over operation of the shop when their father died at the age of 61 in 1936. Before the war started, ALCOA, The Aluminum Company of America, was building a series of hydro-electric dams and power plants in the mountains of Western North Carolina to provide power for their aluminum producing operation in Tennessee. My father and my uncle closed the shop and my father went to work for ALCOA while my uncle went to work for Utah Construction Company who was under contract to ALCOA. My father built a 20' X 20' building near our home and moved the welding machine, a drill press, an arbor grinder and a workbench with a vise into the shop. That original 400 square foot structure is a part of the home that I live in now and that is not inappropriate. Because my life has always been centered around a shop, it is fitting that I live in one now. I can prove all of this because my bare footprint is in the concrete in the front corner of the building with the date, May 9, 1941, written beside it.

I, and later on my younger brother who is three years younger than I, grew up in that shop. We started there as children learning the trades with which we have both made our livelihoods. We accidentally set it on fire one time in about 1955 and we learned a little about fire fighting at that time. Letting the fire go was not an option. Whatever it took to put the fire out was preferable to trying to explain to our father how it got started and as it turned out the damage was minimal. My brother set the house afire in 1945 which was a much more exciting event but that is for another book.

Since that time I have spent a lot of time in a lot of shops. And on construction sites as a weldor. I have spent my time down in the ditch welding the bottom side of the pipe and I have spent time hanging from the boom of a Manitowoc 4000 welding 400 feet in the air. I have made parts for and maintained machinery from offshore oil drilling rigs to rolling mills. I have been in business for myself and failed. I have been in business for myself and succeeded. I have worked for the other man. I have supervised shops and managed shops and I still, at 67 years of age, am happier and more at home in a metal working shop than anywhere else in the world. So it is fitting that I live in a machine shop. And I am incredibly fortunate that I married a woman

Rules, even the good ones, are subject to interpretation. For example, we say that you should never leave the chuck wrench in the chuck. What we really mean is that you should never start a spindle with a chuck wrench in the chuck.

who understands my short comings and who will live with me here in this shop and keep my records and hand me a sandwich or a doughnut through the connecting door now and then. You should all be so lucky!

If it were possible to establish a successful business in the metal working trades simply by reading about it in a book or a magazine or, these days, watching a video or a DVD about it, there would be a machine shop or a welding shop on every corner. I am talking here about the small, independent shop involving the traditional practices of welding, machining, blacksmithing or a combination of all of these. The same is of course true of furniture makers, jewelry makers, auto mechanics, gourmet cooks and chefs or any field of endeavor where the workman and his skills are what is "for sale" so to speak, but I learned long ago to restrict my advice to subjects I know something about. And I know something about metal working. Chiefly, I am beginning to know, at this late stage of my professional life, about all of the things I do not know but I think that is the first step towards real knowledge.

I know that many of you reading this have had someone tell you at one time or another, "Wow! You are really good at this. You should be doing it for a living. Why don't you open up a shop and start your own business?" And it starts you thinking. A friend of mine, gone to his rewards now, and a local sage in my hometown, once said to me, "If you walk from the courthouse to the Post Office (about three blocks for those of you who don't know where I live) and two acquaintances stop you on the same day and say to you that you really should run for sheriff, by the time you get to the post office you are going to have it in your head that you are a shoo-in for the office. And those two votes may be all that you will get on election day."

So think seriously before you decide to give up a lucrative day job and set in to be an independent machine shop operator. Or welding shop operator. Or an artist blacksmith. But it can be done. And I will say to you here and now that if you will follow the rules I am going to lay down here, follow them completely and exactly, I will guarantee that you will be a success! There is of course, as there must always be, a caveat. You must first define success. If success for you means being able to spend four or five months a year at a ski resort, or at the beach, or in a mountain retreat, then you might want to consider alternatives. But if you think of success as I do, and as many of us do, as being able to pay your bills, be warm when it is cold out, cool when it is hot out, and dry when it is raining, then success is achievable. And happiness will follow success if you are doing work you enjoy doing and doing it on your own terms. That is the guarantee. I stand behind it.

There are several things you must have in order to be successful. Some of them require only that you purchase them or build them. Machinery; a place to work; an inventory of materials. These are the easy ones. The more difficult things required are experience, a good reputation, and a clientele upon whom you can depend and who know they can depend on you. These are all components of a successful business and, although some are more readily acquired than others, all are

I seldom accept jobs from people who tell me that they could do the job as well themselves if they had the tools. Maybe they better just get the tools.

Burning, welding and sleeping are all just alike in this way: All three can be done better if you are comfortable.

important. So let's talk about them some. Priorities are not considered here but we will start with what is probably the most difficult and time-consuming to acquire. It is also the most difficult to define.

EXPERIENCE

My brother and I started a fabrication and welding shop business together in 1963 when I was 24 years old and he was 21. We did not succeed in that business and our lack of experience was the primary cause of our failure. I had completed a hitch in the Navy and knew which side of a lathe to stand on and my brother had worked for a year in a boiler shop in Seattle, Washington. We both had grown up learning from our father some of the rudiments of blacksmithing and welding. We enjoyed a ready made reputation among people who knew my father and who remembered my grandfather so we thought we were pretty well prepared to take over the fabrication business in our home town. Wrong! We lasted five years which was probably pretty good, considering, but our age and our inexperience worked against us. We simply did not have the experience we needed to compete for the business which was available in our location at that time. Gaining experience takes time. It takes pain. You have to catch hot slag in your shoe and get hot chips up your nose. It will help if you have crawled into ditches under bulldozers and welded with oil running down the back of your shirt.

And by the way, the mention of hot slag and hot chips brings to mind another thing about the gaining of experience. I don't care how careful you are in working with metals, you will become familiar with the smell of burning hair and flesh and the sight of your own blood. There is no short cut nor is there any way around it. Safety directors and advisors to large companies will tell you that all accidents are preventable and that a zero injury workplace is an attainable goal. I agree to the extent that, in retrospect, all accidents are preventable. I will agree that a zero injury workplace is a worthy goal. But as long as human beings are doing the work and as long as human beings are what they are — accidents are going to happen. It is through experience that you will learn to avoid them and it is through experience that you will learn to take steps to minimize the inevitable.

At some point in your path to gain experience you will begin to think that you are making some headway. That you are finally beginning to learn a little about what it is all about. Nope! You have to borrow money to buy expensive material to meet a deadline for a customer and you will lose that money because the job got screwed up. Maybe your customer was not clear enough in his instructions to you about what he wanted done. Or maybe the truck delivering the finished product to the customer ran off the road somewhere up on the Cowee Mountain and all of your work is now scattered to Hell and gone. Or because — or because — or because — I could fill a thousand pages with things which can happen between getting the job and getting paid for the job and it would only be a start. But whichever of these thousands of reasons causes the loss of the job or

When working on parts which require accuracy to within one ten thousandths of an inch, there must be some sort of climate control. It follows that if the parts are comfortable then the machinist is too. I like that part.

the work, you will be responsible for it and you will be the one who has to do what it takes to do it over. And you will be the one who does not get paid when getting paid is the thing you must have. It is experience which will teach you to be prepared for these things and experience will help you to either avoid them or deal with them when they arise.

Do not confuse experience with elapsed time. We have all known people who think that because they have done the same thing for 40 years that they have 40 years experience when actually what they have is one years experience 40 times over. There is a distinct difference. And elapsed time is not a good yardstick for another reason. It is a fact that some people learn more readily than others do. The child who has to learn that a stove is hot by touching it over and over will likely grow up to be the adult who has to learn the hard lessons over and over. And some will never learn regardless of the manner in which the lesson is presented. It has been said that there is little knowledge to be gained from striking your thumb with a hammer the second time. Most of the point can be gotten across with one demonstration. It hurts! Yet there are people who require the demonstration to be repeated time after time.

There are two basic classifications of experience. What you learn from your own actions and what you learn from the observation of others. Both are good but what you learn from your own actions is probably going to stay with you the longest. It is said by some that you gain experience from making mistakes and you avoid making mistakes by using your experience as a guide. Sort of like the chicken and the egg question, isn't it? But much more complicated. I can enjoy both fried chicken and deviled eggs but the chicken probably is not a fan of either.

Perhaps a good summation of what experience you will need to be successful in a metal-working shop is this:

- You know enough about your work to know there are always alternatives to methods.

- You know to expect the unexpected.

- You realize and take advantage of the fact that you can learn something from everyone, regardless of who they are or what they do.

- You understand that doing things wrong is sometimes a necessary prerequisite to doing them right.

- You have an understanding of the fact that the limitations are a part of the man and not a part of the machine.

- You know your own limitations and how to deal with them.

If you know all of these things you will have a great start towards having the experience necessary to being successful in your business. Remember that the legislative body here is presided over by Murphy, and Murphy's law prevails. If it can go wrong it will. Experience will make you prepared to deal with it when it does.

I wonder why it is that I always seem to produce the best finish on the next to the last pass.

Out of position welding for me used to be an overhead weld in a gravel pit. Now it is out of position if it isn't lying right there in front of me at waist height on a welding table. With a chair to sit in.

A SHOP

A good next logical step after you acquire the EXPERIENCE will be a place to work or a SHOP. This is what it is all about isn't it? Having your own SHOP. It must be large enough to accommodate you, your machinery and the accumulation of scrap material, useless fasteners, machine carcasses and actual useful inventory that seems to go with shops. It must be suitable to the climate in which you live, capable of being heated if you live in Buffalo, New York, and capable of being air-conditioned if you live in Houston, Texas. It must have large doors if you work on large parts, a loading dock if you have to ship and receive materials, a floor suitable for moving and installing heavy machines if your machines are heavy and some means of lifting and moving those machines.

When your customer assures you that it doesn't matter what kind of steel you use to make his part — it matters. And when the part fails it will be your fault.

Above all, it must be suited to what you are doing. I have been asked many times, "How thick should the concrete be for a good shop floor?" Or, "How high should the ceilings be in a machine shop?" Or, "What is a good size for a shop to be?" And there are no uniform answers to those questions. If you are going to install punch presses, even small ones, your floor will need to be heavy enough to isolate the pounding from precision machines. If you plan to install vertical machining centers which are ten feet tall then you will have to have enough ceiling to accommodate them and don't forget to build a door large enough to get them inside. I know, this sounds elemental. But I am continually amazed at how many people will build, sometimes at considerable expense, a shop or a facility which doesn't meet even their basic needs and then they sit and cry about it. Build your shop to accommodate what you plan to do in your shop! How difficult can that be?

The only truly dumb question is the one which should have been asked but was not. But I have heard some asked that come pretty close to being dumb.

And build for the future. When I retired from ALCOA in 1993 my intent was to go back into business, at least on a part time basis, doing work I had done back in the 1960s and 1970s. And that was being an artist blacksmith. I like the work. There is a large and growing market now for really good ornamental iron work and I live in an area where large houses with lots of hand railing and big fireplaces are being built all over the mountains. I would discourage that if I could but that is a matter for me to take up with the county government. So I prepared myself and my shop to begin building fireplace screens and hand forged iron furniture. Within a year I had found that the market for a reasonably talented machinist was even better than that for a blacksmith. (The fact that the work is less physically demanding had nothing to do with this decision, of course!) So I sold my 14" lathe and bought a 17" lathe and began filling my shop with machine tools which now includes a CNC lathe. I am relating this here to point out this simple fact. What you think you will be doing in your shop will not necessarily be what you will actually do.

So when you build, or buy, or rent the place where you will have your shop keep all of the options open that you possibly can. Be prepared to jump in whichever direction the ball is coming from . Sometimes you will be limited and you will have to get along with less than the optimal requirements for a place to work but here is a good rule to go by. The size of shop you will require is approximately 10 percent greater than the size you will have. Or put another way, you will accumulate machines and inventory at a 10 percent greater rate than you will be able to provide space for.

EMPLOYEES

After your shop is built and the work starts coming in you will begin to think about HIRED HELP or EMPLOYEES. My deepest sympathies go out to you at this point. Moderately sized shops and even some very small shops need, and can benefit from, having the right kind of help. It is true, however, that of all the pitfalls of being in business for yourself this one has the potential for being the deepest. If you really need help then there is no option. You need help. But where do you go to get it? Sometimes it really is so that your brother-in-law can learn the work. Or that he can already do the work. I guess I should say either your brother-in-law or your sister-in-law but, at the risk of offending the equality-of-the-sexes people, I have to say that I am old school and I still think of machine shops as being filled with men who go home to be fed by women. I know I am wrong! But I am old and I grew up in a different time.

To get back to the point of the discussion, where do you go to find qualified help. A good source is a trade school if you are fortunate enough to have one located close to you. Instructors in these schools will be happy to advise you about the capability of their students and will help you because it helps them. Any interested young person is a possibility whether from a trade school or not but they must be trained. And here you will be making an investment which is like all investments. You take a chance. It can be discouraging to spend the time it takes to train a promising young person to do the work as you like for it to be done and then have them leave taking all of the time you have spent on them with them. If I had the answer I would be a rich man by now but I don't and I don't know who does.

Hiring itinerant machinists or weldors has its risks as well. Metal working shops, machine shops in particular, require as high an investment in terms of equipment as almost any field of work in the country and when you hire a person off the street to run your $75,000 machining center just because he says he can do it you may sometimes find yourself wondering what those loud noises, which sound suspiciously like violent train wrecks, could possibly be.

It should go without saying that any unproven hired help should be required to work under the supervision of either yourself or someone whose judgment you can trust for a period of time before turning them loose with your machinery. It has always been and will always be a source of resentment to me the way weldors and pipe fitters are hired on many constructions jobs. When a pipe fitter went to the employment office and applied for a job he was asked the usual questions. "How much experience have you had?" "Where did you work last?" And so on. But when a weldor went to the same office and applied he was told to go to the test shop and prove that he could weld. Yet on most

A sample part beats just a description when you are asked to make a part. But a good drawing is by far the best.

jobs they were paid the same wage. It just never seemed fair to me.

And don't forget about the liabilities of hiring people. I once hired a young man, a beginning machinist with some good background and a man certainly capable of learning. He chucked a part up in an engine lathe and, through simply not paying attention, he started the spindle at a speed of 2000 RPM when he intended to start it at 490 RPM. He only shifted one of the speed control levers and should have shifted two! The part, which was a piece of CRS weighing approximately 34 pounds, was flung violently out of the lathe with enough force to knock the headstock out of alignment. Fortunately the damage to the machine was negligible and was easily repaired. The operator himself suffered no physical harm. And it was not intentional, it was just a mistake. But I told him the next day after sleeping on it overnight that although I knew it was an unintentional mistake and I make mistakes myself, this one cost him his job. I could not sleep with the knowledge that a simple mistake might make his widow the owner of my shop! I understand that there are insurances available for liabilities of this nature. And if you have enough work to justify hiring employees you will have to have this insurance. Just remember that if a man is working for you then everything he does will have your name on it and remember too that if he screws up you are the one who will have to pay for it.

How can you really appreciate a good day if you don't have a bad one every now and then?

CUSTOMERS

Just as important as having a SHOP is having CUSTOMERS. If you don't have customers you will certainly have lots of time to engage in machine work as a hobby but you better have either a good day job or wealthy parents because customers pay the bills and metal working machinery and materials generate lots of bills.

CUSTOMERS, bless their hearts, are a mixed bag. I once heard a wise man observe about his business, which had nothing at all to do with metal working, that, "If it weren't for the help and the customers this would be a great way to make a living." And I could sympathize with him. When my brother and I started our shop in 1963 we had an oxy-acetylene torch, a couple of welding machines, a drill press and a bench vise. We needed customers and we needed them so badly that we had to accept, or we thought we had to accept, every job that came in the door. If someone came in with a job and we knew that this person had a bad reputation for paying his bills — we took the job anyway. If a job came in that was dirty or difficult or dangerous or even illegal — we took it! We knew that our livelihood depended upon our ability to get jobs first and our ability to do the job second so we took on every CUSTOMER whose shadow darkened the door and sometimes we paid the price.

It seems the less experience a machinist has the closer the tolerances to which he claims to work. I wonder why that is.

If we only take the jobs that CUSTOMERS bring us that we know are lucrative, that we know have little or no risk, or that everything about the job will be fun to do, then I really believe we will be spending a lot of time standing in the doorway, looking up and down the street and wondering what we will be doing next. In many ways the problems that CUSTOMERS bring us will reflect directly upon what our EXPERIENCE has taught us. You can only recognize a bad customer by having had dealings with him or his counterpart earlier. And you can't tell the difference between good CUSTOMERS and bad CUSTOMERS unless you have had some dealings with both.

You will never have a second opportunity to make a first impression. So you must always be prepared for the time when a potentially good CUSTOMER will walk in. If the representative of a multi-million dollar manufacturer drops in unannounced and sees your shop in a mess or sees you or your employees blatantly ignoring safety practices or any other violation of good shop practice, then you lose the opportunity at that time of possibly gaining what could be a CUSTOMER who could guarantee your success. Sometimes it takes no more than that to make you a success or a failure.

I will only add here that when you acquire a good CUSTOMER you have to show him your appreciation. You have to make him or her want to come back to your shop over and over because he or she knows that from you they will receive immediate attention, good quality service and dependable delivery. And the proof is in the pudding.

MACHINERY

This is the easy part. And the fun part. The MACHINERY. In my shop there are two basic types of MACHINERY. Those pieces that I needed at the time of their purchase and those that I merely wanted. The two types are not necessarily the same. We can get pretty inventive when it comes to the justification for a particular piece of machinery and if all of you reading this are like me the only piece of machinery you don't want is the one you don't have. But when you are doing this for a living there are some concessions to common sense that must be made. Let us take for example an iron worker. The machine — not the man. This is something I do not have in my shop although I did own a small hydraulic machine at one point. What I do for my customers is mostly pretty basic machine work. Turning, milling and grinding with some welding and heat treating thrown in. Generally it is work which does not require the things that an iron worker is designed to do. But I frequently will do jobs for myself, building a new hoist out at the back door for example, which will require the fabrication of several gusset plates from ¼" steel. So I want an iron worker. I don't need it because I can saw or burn the gussets out. I can drill any holes required in them on the drill press. But if I had an iron worker I could shear the parts and punch the holes. So I want an iron worker. And if I am completely honest with you here, one day I will probably have one even though the nature of the work that I am mostly called on to do for my customers does not require it.

MACHINERY is what we have to do what our CUSTOMERS want us to do for them. If you go into this business just to do what you have to do then you may as well keep the job you are already working at. Most of us in this business are doing these things because we want to do them and the MACHINERY that

If the jobs are not complicated and the machines are close to each other and the lunar signs are just right — then it is OK to run more than one conventional machine tool at the same time. But be careful!

we have sought out and purchased is what enables us to do it.

As far as the practical aspect of recommending what machinery you will need in your business I cannot do that. And you don't expect me to. You need the machines to do the work that you have chosen to do. If that work calls for EDM machines then by all means buy EDM machines. If all you intend to do is build and install hand railings or fire screens then likely all you will need will be a MIG welder and a drill press. And an iron worker. It seems pointless to say this but really all that is required is for you to have the MACHINERY to do what your CUSTOMERS have hired you to do.

I will say this, however. It seems there are two distinct philosophies to acquiring machinery. The first is this. You decide what type of work you are going to go looking for and you then go out and find, and purchase, the machinery to do the job. And then you look around for the job. That is my method. I hesitate to tell potential customers that I can do this process or that process in my shop unless the machinery is in place and hooked up and ready to go. The other way is go find the job and then get the machinery needed to do it. I will admit that the latter method is probably the best method. Larger shops, and even some small shops, will hire a manufacturer's representative to go out and sell their services. If the salesman does his job you will sometimes be caught in the awkward position of having a job, complete with a required delivery date, and not have the means at hand to do the job. But practically speaking, all you have to do is buy the machinery. Jobs and customers are harder in the long run to come by.

Where you acquire your machinery is filled with options. If you need specific machines for specific purposes you will be limited sometimes to either purchasing new machinery or, if you are really lucky, finding the machine you are looking for on the used market. But if you just need a lathe, for example, you have several choices. You can buy a new one. And if you know that you will be depending upon the machine every day perhaps buying a new one is what you should do.

Do your research as to what the machine you are considering buying will do compared to your requirements. Consider the potential growth of the market you will be entering with this new lathe. Can you assure yourself that the 6" diameter X 14" length of the parts you need to make on this new machine will be the largest parts you will ever be required to make? Or should you consider buying a larger machine so that you can go after the other job which might be just over the horizon?

One of the things it seems to me is most often overlooked by potential machine tool buyers is the capacity of the machine. And what I call the false economy of buying based solely on purchase price. I preach to every congregation who will listen that you don't save money by buying cheap tools. And the larger the purchase, and/or the perceived "savings" to be realized by buying only the bargain basement tools, the greater potential there is for disappointment later on. To continue with our example of buying a lathe to machine a 6" X 14" part. You have found a used lathe

We like what we do and we will work for free. But we charge $65 an hour for the use of the machines.

There will come a time when the machine is worn out. Sometimes it will get emotional and you have my sympathy. Shed a few tears if you must but get rid of it, buy a new one and get on with your life.

for sale by a widow whose husband owned a small machine shop in a town just 40 miles away from where you are. The machine is clean and seems to be well cared for. There is a nice quick change tool post with it. It has a couple of chucks and a collet closer and you can buy the machine for $2500. It has a 1 HP motor, single phase which you require since you do not have 3 phase power at your shop. And it is a 12" X 40" machine so you can chuck up and turn the 6" X 14" part your customer wants.

On the other hand there is another choice, a used 16" X 48" machine another 130 miles down the road. It has comparable tooling and has also been well cared for and is in good condition. This one has a 3 HP, 3 phase motor on it and will cost you an additional $1000 to buy plus you will have to find a bigger trailer to bring it home on and you will have to hire somebody with a crane or a forklift to unload it and put it into your shop. And you will have to install some sort of phase converter in order to run it. In total this machine is going to cost you about twice as much to buy and put into service as the 12" machine. So what do you do? Both machines will accept and machine the part and money doesn't grow on trees.

If you know that you will never have a bigger part to machine and you know that you will never want to buy another machine which requires 3 phase power and if you are satisfied with having to never exceed taking a .08" depth of cut at .004" IPR feed rate then buy the smaller machine and put the balance in the bank. But the smart machine to buy in this case is the bigger machine. You will spend the money for the phase converter and the larger lathe one time and you will use the machine every day. If you are serious about being in the machine shop business then be serious about it.

This example is maybe not a good one these days when the choices we have to make are not between spending $2500 and $5000 but rather tend to be more in the range of ten times that amount. But the point I am trying to make here is this. Don't be near-sighted in buying your MACHINERY. None of us have crystal balls, at least nobody of my acquaintance, but when you are preparing to buy a machine tool look not only at what that tool will be expected to do this week but what it will do, or might do, next week or next month or next year.

And you might get lucky and find a widow who is trying to sell just exactly the machine or machines you are looking for. It has happened — but never to me.

Other things to consider in your decision to go into the metal working business are PARTNERS, a LOCATION, and a REPUTATION.

I have no advice on PARTNERS. My own PARTNER also keeps my books and washes my clothes and does the cooking. She has been a PARTNER to me now for 43 years and I have about decided that the trial period is over and she now has tenure.

LOCATION will depend upon what sort of clientele you plan to cultivate. If you want to get all of the broken lawn mower handles and trailer hitch

CNC machine tools are wonderful. They can generate tool paths and produce profiles never before possible. But the best thing about them is when you roll that door shut the chips and the coolant are on one side and you are on the other.

*A weldor calls a 5/32"
welding electrode a five-
thirty two. To a machin-
ist that is a number 5
screw with 32 threads
per inch. That prob-
ably is not important.*

installations then you need to be in a highly visible, convenient location. I am not being critical of that part of the business. Many small repair shops have built a success by doing just that sort of work and there is no denying that the business will always be there. Our society is becoming more and more of a "throwaway" society but there will always be people who would rather have the old one fixed instead of replacing it. If your business is one which depends more upon pre-arranged contract work or if you manufacture a product and ship it, then your LOCATION only needs to be convenient to FEDEX or UPS or to whoever you use to ship your product.

A REPUTATION is something you will have to take care of for yourself. It is a certainty that you will acquire a REPUTATION but it is up to you what form it will take. If you get a REPUTATION for doing accurate, high quality machine work then that REPUTATION will grow and spread. If your REPUTATION is one of a weldor who uses the cheapest materials available and who does just enough to get by on a job then that REPUTATION will also grow and spread. A REPUTATION is the one thing of all the facets of being in business which will take care of itself. All you have to do is get it started and head it in the right direction.

All of the things we have discussed here are important. Some will have greater or lesser importance to you depending upon what direction your business venture takes. As I stated earlier I am already past the Social Security barrier so most of my work has already been done. But Nan knows, because I have told her, that I intend to run my lathe or my milling machine as long as she will wheel me up to it on a hand truck. This is what I do. I consider myself among the most fortunate of men because what I do for a living I have always done for fun. I have worked for the other man. I have failed in a business venture. And now I am successful beyond all I ever dreamed of because I have defined success. I am doing exactly what I want to do and I am getting paid for it. I have known people and I know some of them now who are miserable in their jobs but they keep doing it because they think they have to. If I am fortunate enough to survive to an age where the young news reporter comes to me and says, "Tell me, Mr. Bulgin. What is your secret for longevity?" I will tell him that the secret is this. That what you are doing is only work if you had rather be doing something else. As for me — there is nothing else I would rather do nor is there any place I would rather be than playing in my shop.

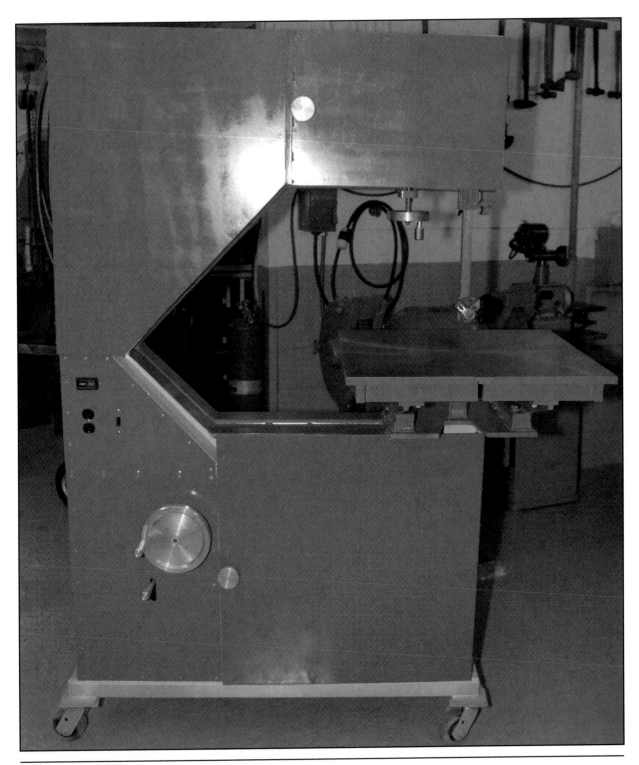

Chapter 2

Building a Variable Speed
Vertical Bandsaw

Introduction

Well, here it is. This is the chapter that I have been talking about and emailing about and bragging about for about three years now. Or more. Building a variable speed vertical band saw. I built the original saw, shown here in Photo A, in 1978, and it has seen thousands of hours of use and is still a good, no, make that a great, vertical band saw. The second machine, the one I have built for this article, can never be any better than the first one and I only hope that it is as good. I made a few changes to take advantage of the hindsight I gained from building the first one but it has been so long since that time that I made most of the mistakes again. The only difference is this time I have a photographic record of my screw-ups!

Producing a good finish on a part is not difficult. Machining parts to exacting dimensions is just a matter of patience. But getting exactly the finish required at precisely the correct size takes some experience.

This is not just a hobby project. This is a serious machine tool and I will tell you now that the time required to build this machine is not insignificant. If you have the time to build it you will not be sorry but if you just need to have a good bandsaw for your shop, and if your time is really valuable, then you may be better served by buying a good used bandsaw like a DoAll® or a Grob® or any of several other machines available today including some from "over there." But then you would not have the satisfaction of showing people through your shop and pointing to your band saw and saying to them, "This is one of my machine tools I use the most. I built it myself." And that is a really satisfying thing to be able to do.

A few notes before we get on with the job at hand. This machine is of three wheeled design, all three wheels being of the same diameter. You may wish to go about it differently. You can make the third, or offset, wheel of a smaller diameter than the other two, or add a fourth wheel, which can give you a deeper throat with basically the same framework. If you have a lathe of large enough capacity to make larger wheels you can eliminate the third wheel and make a two wheeled saw. Another option would be the simplification of the drive train, shown here, but then you will limit yourself to sawing only wood or only metal and you would lose the chief advantage of a variable speed saw. There are many other opportunities for changing the basic design which is illustrated here and I will be surprised if many of you do not make changes and improvements. All I ask is that if you come up with important changes which improve the basic machine is that you tell me about it so that I can make my machine better.

I am providing detailed drawings and directions with this chapter for the construction of the frame, the drive mechanism and the wheels and table. If you follow the directions and the information given in the drawings exactly, step by step, you will have a saw just like mine, which isn't a bad thing. But I am sure that many of you will regard these directions as they are intended, as guides rather than as strict directions. For example, I am of average height, about 5' 11" tall. If you are of shorter or taller stature then the table height should be such that it is comfortable for you to use.

A change of only an inch or so can make a huge difference when you are standing at the machine for extended lengths of time. I saw a machine once which had been modified so that a paraplegic could use it from a wheel chair. An extreme example, to be sure, but just one more example of what can be done if you are building a machine for yourself or for someone who you know has different needs from the average. Another example of this is that I have never seen a machine designed for a left-handed operator. That is, with the switch and the speed controls on the right. Not a big thing, I will grant you, but something to be considered nevertheless. With these plans all you have to do is get a mirror to view the drawings! Or maybe not.

A. The original saw built in 1978.

It is inevitable, at least for me, in a design-as-you-go job of this nature, that there will appear some contradictions between what you see in the drawings and the photos. Or maybe from one photo to another appearing later on. I will try to address those changes as we go. In most cases they will not be significant, but I suggest that, if you plan to build this bandsaw, you read through the text completely at first and take note of those differences. An example of what I am talking about may be seen in the installation of the table mounting plate, or part number P-11. In the photos of the welding process you will see the plate welded into place without the notch for blade clearance. The isometric drawings, drawing numbers 4 and 5, of the frame do not show it. Nor do the photos show the gusset plates welded on under the table mounting plate. But if you will look at the detail drawing for part number P-11 you will see both the size and location of the notch and the position of the plate relative to the frame. I hope there won't be too many of these and that they will not be a major inconvenience to you.

The material lists provided here are for building exactly the machine you see in these photos and drawings. I used aluminum for the wheels and steel for the pulleys because I had the material and it works well. But steel wheels and aluminum pulleys or a combination of both will work as well, and you can use either solid wheels and pulleys or fabricate them. They just need to be round! If you have or can find a used cast iron table top which will work dimensionally for you then that would be a good option. Cast iron makes a better table than the steel one I made because the friction between the work and the table will be reduced to a noticeable degree, especially for the first several hours of operation. The Blanchard finish on the table shown here is a good second best.

This machine is built heavier and more robust in places than is necessary for dependability and continued use. There are some reasons for that. I have used in most cases materials that were on hand in my shop, materials either left over from jobs I have done for customers over the years or purchased in one of the many auctions I have been to. What has been economical for me because of the ready availability of materials, even if they are heavier than is necessary to do the job, may cost more than you will want to put into the project. Another reason for heavy materials is that sometimes fabrication methods can dictate material choices. The table and its mounting structure is an example. If the table top were to be made from 1/4" plate, using the welding processes available to me, the degree of distortion from welding would be such that grinding the finished part on a Blanchard grinder would make a problem for the man running the grinding machine. And there are other examples throughout the entire job. So you can make some choices and some substitutions. This machine in my shop, and I will assume in most shops where it will be used, will be set into place, perhaps bolted to the floor, and put to work. Its added weight will not be a disadvantage. And you will hopefully appreciate, as I have, the fact that maintaining the machine over the years has been limited to replacing the blades from time to time. I did have to replace one of the heater strips in the motor starter back in 1994.

I am foregoing the opportunity here in these instructions to give good advice about basic fabrication and machining practices. Keeping your work square, cleaning the welds, welding in selective areas to avoid distortion and all that stuff. I am assuming that anyone who sets out to build this machine will have what it takes in terms of welding expertise, machining equipment, experience, etc. In fact, there will probably be some of you building the saw who can teach me some things about it and as I think I have already said, I will appreciate hearing about it. One thing you will need in the fabrication process, unless you are 9 feet tall and a weight lifter, is some sort of lifting device. You can see the ones I used on this job in several of the photos here.

There are some good things to be said about reaching retirement age but I will tell you that it ain't all gravy!

One more note: I would hate to set out to build one of these machines without already having one to use for the job. I don't know how you will get around that one.

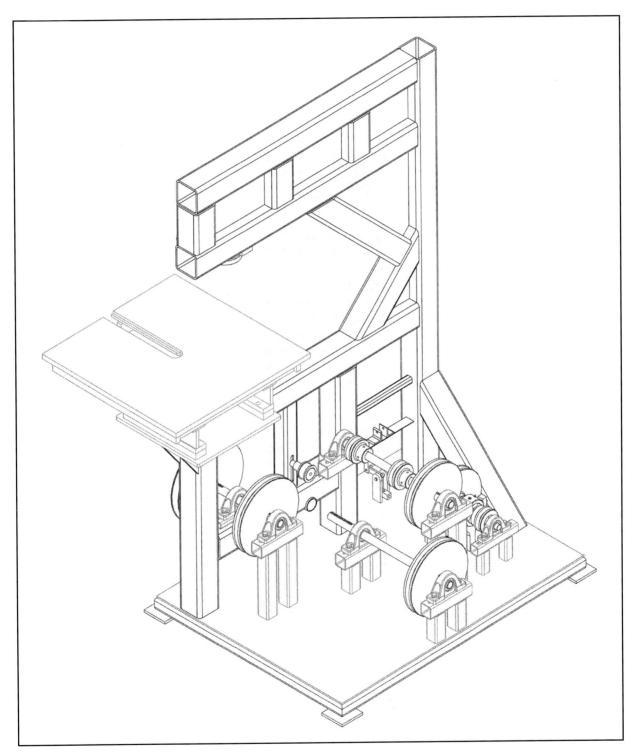

Isometric Drawing No. 1.

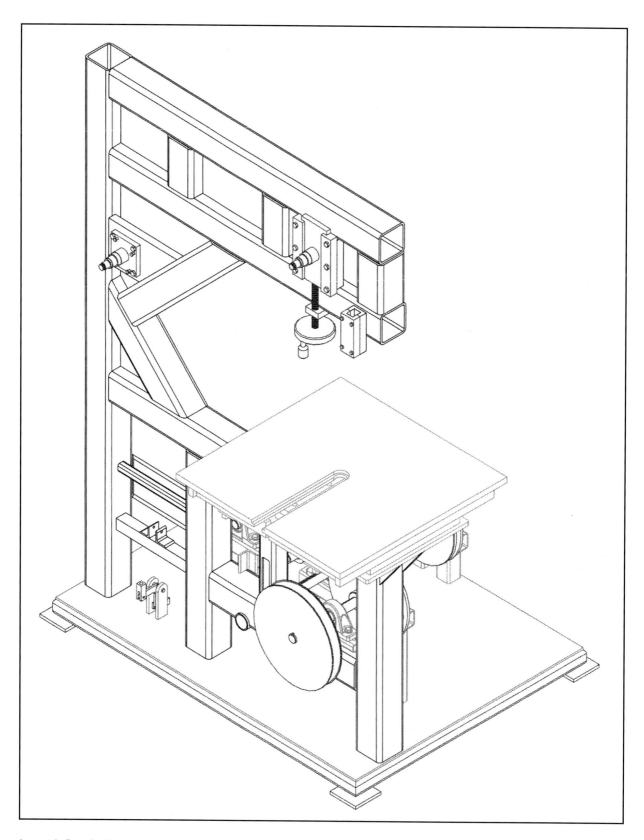

Isometric Drawing No. 2.

Frame and base, welding complete.

I once had a customer who would bring things to me to be repaired that the junk man would turn away. I once told him that he needed to learn to recognize when something was worn out.

SECTION ONE: Building the Base and Frame

Begin the construction by fabricating the base from the 2" square tubing and the 1/4" plate as shown in Drawing No. 1. It is also shown, incomplete, in Photo 1. The bracing underneath the floor plate does not need to be welded 100% but be sure to weld the points directly underneath the areas where the vertical frame will be attached. Complete the welding of the underside before you turn the plate over including welding on the mounting pads at the corners. If you have planned well this may well be the last time you will ever see the bottom of the machine. The casters you see in several of the photos were only used during the construction process. They made it much easier to move the thing (it will be just a "thing" for a while yet!) around in the shop. If you do want to

MATERIALS LIST 1
Base and Frame

1. 2" square tubing — 1/4"-wall thickness. A 20' length will make the base and you will have some left over for miscellaneous items.

2. 1/4" plate — 1 piece 48" X 36" for the base. It will take about an additional 3-square-feet of plate, either 1/4" or 3/16", for the filler gussets in the upper frame.

3. 4" square tubing — 1/4" wall thickness. You will need the best part of (2) 20' lengths of this. It takes about 35' total to build the frame.

4. 1/2" plate — 1 piece 16" X 16" for the table mounting plate. There are also a couple of gussets supporting this plate which will take another 4-1/2" X 6" piece of material.

5. Pulleys — (2) small pulleys 1" to 1-1/2" in diameter. These are for the blade guide counterweight. Get good ones, preferably with ball bearings, because they will be installed where maintenance could be a problem.

6. (2) pieces of 1" X 2" HRS 14" long. Table mount attachment.

7. 3/4" or 1" steel plate — 1 piece 6" X 10" and one piece 6" X 8". Spindle mounting points for the upper wheels.

8. 3/8" X 4" X 4" HRS — (4) pieces for mounting pads.

install permanent casters then change your frame dimensions accordingly so that the table height and other critical elevations will not be affected.

You may choose to lay the main frame out on the floor and get it tacked up before welding it to the base. It is a little easier to maintain squareness and, if your floor is smooth and level, you can keep everything in the same plane. My welding table is a piece of 2" A36 plate 4' X 9', so that is where I built this one. And it gave me a good reason to clean off the welding table which I try to do at least every 6 or 8 months. Drawing No. 2 is a front elevation which gives the basic dimensions for the frame and Drawing No. 3 is of some sectional views which should clarify some of the details. Photo 2 is of a bald headed weldor that I sometimes use for the more menial tasks around my shop. Actually, he is a retired Level 3 Radiographer and welding inspector who was also a certified nuclear weldor at one time. In real life he is my younger brother, Fred. Photos 3, 4 and 5 are other views of the fabrication process for the machine and Photo 6 is of the completed and primed frame. There will be other welding to do as we proceed but I live in a pretty humid climate and the interim coat of primer helps to keep the rust down until I get ready to paint the finished machine.

Install the small pulleys inside the tubing at this time. They may be installed at a later stage of the construction but it is much easier to put them in place now. The sole purpose for these pulleys is to support the counterweight which will be installed later to offset the weight of the upper blade guide assembly. If you use good pulleys here and use locking nuts on their axles there should

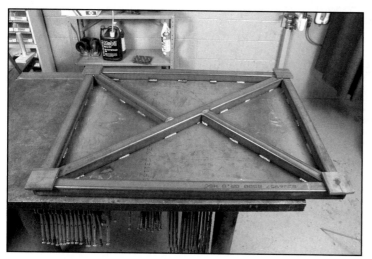

1. Bottom side of machine base being assembled.

2. Welding using the MIG or GMAW process.

be no need to ever have to access them for maintenance. The end caps which are welded on the ends of the 4" tubing for the finishing effect will be welded with the minimum amount of weld to keep them in place so that they could be removed should there ever be a need. See Photos 7 and 8 for reference. In Photo 7 the upper blade guide mounting bar is in place showing the relationship of the pulleys to each other.

3. Partially assembled frame.

The two isometric drawings, Drawing Nos. 4 and 5, are included here for clarification purposes only. The exact configuration and positioning of the table mounting plate will be shown later when we fabricate the table itself.

The mounting pedestals for the bearings and the other components of the shifting mechanism will be installed in Section Two.

4. Another view of the frame.

5. Frame welding complete.

6. Frame welded and primed.

7. Counterweight pulleys with upper blade guide support in place.

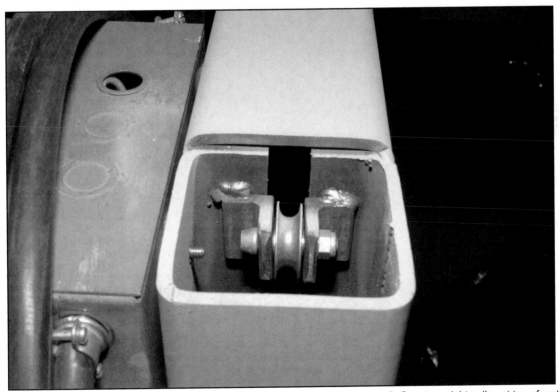

8. Counterweight pulley at top of vertical column.

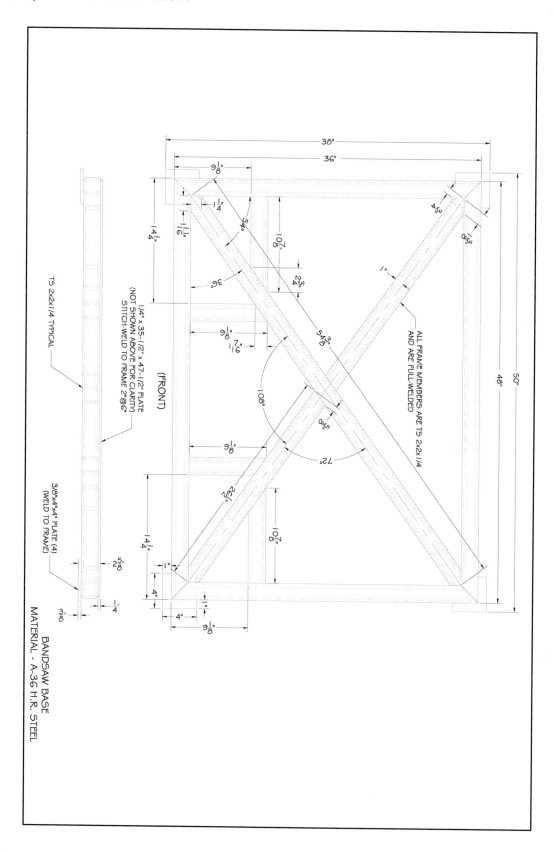

Drawing No. 1.

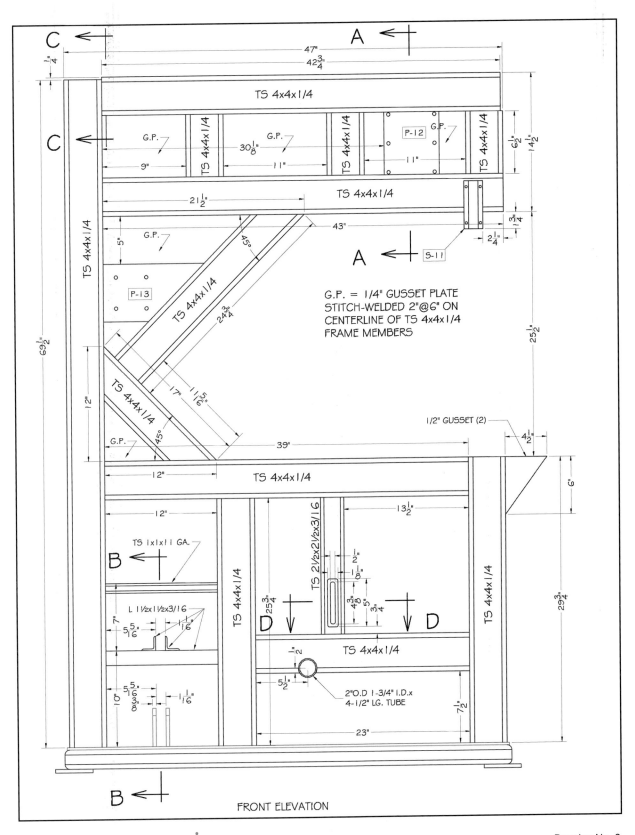

FRONT ELEVATION

Drawing No. 2.

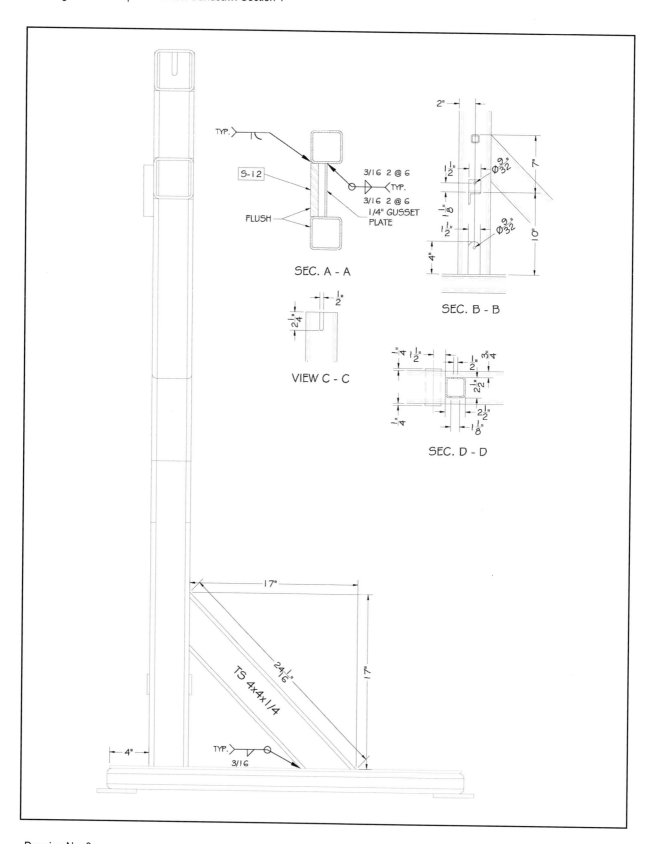

TYP.

S-12

FLUSH

3/16 2 @ 6
TYP.
3/16 2 @ 6
1/4" GUSSET
PLATE

SEC. A - A

VIEW C - C

2"

SEC. B - B

SEC. D - D

17"

17"

TS 4x4x1/4

24 1/6"

4"

TYP.
3/16

Drawing No. 3

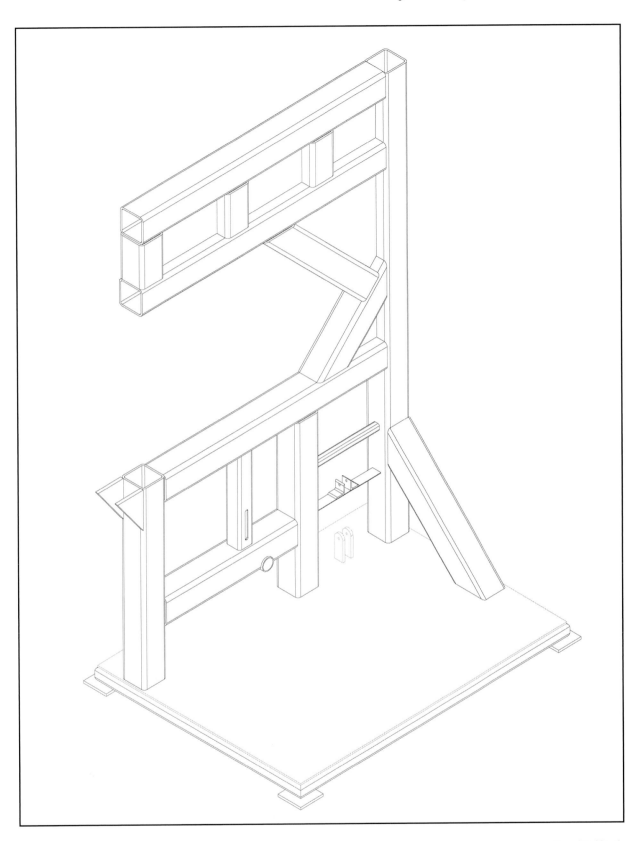

Drawing No. 4.

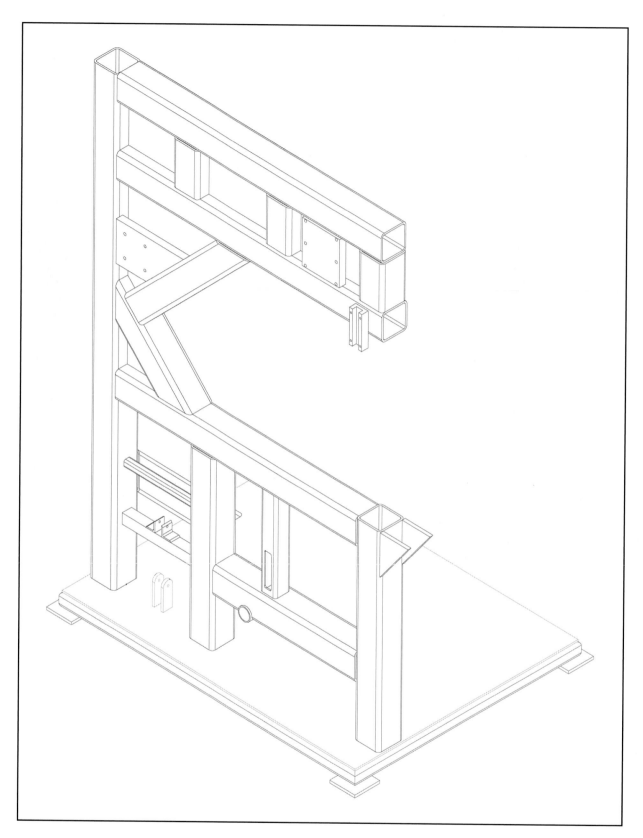

Drawing No. 5.

Pulleys in place.

SECTION TWO: Variable Speed Drive

I know that I am going to get letters and emails and maybe even telephone calls and telegrams about the weight and dimensions of the pulleys I use in this drive system so let me see if I can head them off. I machine a lot of steel and aluminum discs for a customer and I order the discs precut from my material suppliers. The aluminum is either sawed from plate or sliced from bar stock. My supplier keeps aluminum bar stock in diameters up to 14." The steel discs are either burned from plate or they may also be sawed from a bar. I had surplus material on hand so I used it. As I have said many times and will say many times more, material selection for jobs which I am doing for myself are influenced partly by its suitability for the job, partly for its ability to be fabricated readily, but mostly because it is already in my shop and is surplus to other needs. I do not have a scrap box in my shop. I have instead a secondary material storage bin. I consider chips and swarf to be all the scrap I produce.

I made most of these pulleys from 1-1/2" HRS plate. 1" plate would certainly suffice and if you choose to run A size V-belts instead of the B

"I'll just order a new one when I need it."
— Not a good way to manage your supply of fire extinguishers.

size I have used, then ¾" plate will serve. The 3" diameter steel hub welded onto the pulleys is a useful design feature for two reasons. On two of the pulleys it is needed in the Hi-Lo drive where the driving lug engages but it also serves well as a point to chuck on while machining the V grooves and the bore. The groove and the bore must run concentric with each other and this welded-on hub makes it possible to machine both in one setup. Other possibilities for pulleys are fabricated pulleys, spoked pulleys, cast pulleys or pulleys purchased from McMaster/Carr, Grainger or some other industrial supplier. But remember that if you change the diameter of one pulley you must do the arithmetic to change all of the other pulleys between that point and the blade of the saw. This reminder also applies to motor speeds.

One additional comment about the heavy pulleys. Although they are far heavier than is really necessary to do the job, the inertia of the rotating mass while the saw is in use makes for a really smooth

Description of pulleys in this project

PULLEY	DIAMETER	GROOVES	BELT	BORE SIZE	NOTES	RPM
No. 1	3.69"	single	B size	to fit motor shaft		1750
No. 2	11.17"	single	B size	1-1/8" with 1/4" keyway		578.3
No. 3	8-1-C Driver pulley from Speed Selector Inc.			1-1/8" with 1/4" keyway	common shaft with Pulley No. 2	578.3
No. 4	8-1-M Driven pulley from Speed Selector Inc.			1-1/8" with 1/4" keyway		213.2 thru 1599
No. 5	3.6"	single	B size	mounted on ball bearing so it can turn freely on shaft when Hi-Lo dog is disengaged	common shaft with pulley No. 4	213.2 through 1599
No. 6	9.9"	single	B size	mounted on ball bearing so it can turn freely on shaft when Hi-Lo dog is disengaged	common shaft with Pulley No. 4	213.2 thru 1599
No. 7	3.6"	single	B size	1-1/8" with 1/4" keyway	common shaft with Pulleys 8 and 9	77.5 thru 579.2 in low range/ 586.2 thru 4380.5 in high range
No. 8	9.9"	single	B size	1-1/8" with 1/4" keyway	common shaft with Pulleys 7 and 9	77.5 thru 579.2 in low range/586.2 through 4380.5 in high range
No. 9	3.83"	single	B size	1-1/8" with 1/4" keyway	common shaft with Pulleys 7 and 8	77.5 thru 579.2 in low range/586.2 through 4380.5 in high range
No. 10	11.52"	single	B-size	1-1/8" with 1/4" keyway	common shaft with bottom saw wheel	25.6 thru 192.6 in low/ 194.9 through 1456.5 in high range

running machine. The balance is maintained by machining the pulleys all over during their construction and the results will be obvious.

The table lists the pulleys shown in the accompanying drawings.

This arrangement of pulleys with the given sizes will, in a perfect world, give you from 83 FPM through 4718 FPM blade speed with a slight gap between 624 and 631 FPM. This allows blade speeds for nearly all materials which lend themselves to being sawed on a vertical bandsaw.

The ball bearings used were bearings on hand at the time of construction. The bearings included in the material list are of good dimensions and load bearing capability for the job, but substitutions can readily be made. If you do not use sealed bearings here you should make provisions for installing separate seals because of the location and environment. Using sealed bearings is a lot easier.

The material list shows everything required for building a drive exactly like the one described. Your materials may differ slightly or they may differ radically from the list shown here. You will, of course, need a motor, which I do not list here because there are so many variables. The motor on my original saw is 1-1/2 HP. The motor I used on this saw is a GE 1-HP motor. The basic requirement is the speed which should be 1750 RPM. A 3/4-HP motor might work but I would prefer to use at least 1 HP.

Begin building the drive mechanism by first doing all of the machining. This includes the pulleys, the shafting and all of the components of the Hi-Lo shaft and its associated shifting mechanism. The pulleys themselves take the most time to complete so that is where I started. Drawings 6 through 11 provide details of all of the pulleys. The larger pulleys, nos. 2, 6, 8 and 10, are of two piece construction. Machine a locating recess in one side of each of the pieces of 1-1/2" plate (or whatever thickness you may want to substitute) for these four large pulleys. Photograph 12 shows how this is used to locate the hub and how it also provides a welding prep for welding on the hub. Photograph 13 shows the welding process. I am welding here on the new burning/welding table I built for Chapter 16. Note that the grounding clamp is attached to the table top and not to the base in order to make sure you do not cause an arc in the bearing. One heavy pass with a 1/8"

9. Top, homemade variable speed pulley open.
10. Center, homemade variable speed pulley closed.
11. Bottom, fixture for machining variable speed pulleys.

MATERIAL LIST
Variable Speed Drive

1. HRS or CRS steel plate circles, burned or sawed to the following sizes:
   (1) piece 3-3/4" round X 3" — Pulley 1
   (1) piece 11-1/4" round X 1-1/2" — Pulley 2
   (2) pieces 3-3/4" round X 3" — Pulleys 5 and 7
   (2) pieces 10-1/8" round X 1-1/2" — Pulleys 6 and 8
   (1) piece 4" round X 3" — Pulley 9
   (1) piece 11-3/4" diameter X 1-1/2" — Pulley 10
   (1) piece 2-1/2" diameter X 1-1/2" — idler
   (4) pieces 3" diameter X 1-5/8" — hubs for pulleys 2, 6, 8, and 10
   (1) piece 3" diameter X 3" — sliding dog for Hi-Lo shaft

2. CRS shafting — 6 feet of 1-1/8" shaft will make all of the plain shafts in the lower housing. You will also need about a foot of 3/4" shafting for the speed control handwheel.

3. 1045 TGP shafting — One piece 1-7/16" diameter X 15-1/4" long for Hi-Lo shaft.

4. (8) 1-1/8" pillow block ball bearings
   (2) 3/4" pillow block bearings — bronze sleeved pillow blocks will work OK here.
   (1) 5205 2RS bearing for pulley #5
   (1) 6307 2RS bearing for pulley #6
   (1) 5204 2RS bearing for belt tensioning idler

5. (1) set of variable speed pulleys from:　　　Speed Selector, Inc.
   17050 Munn Road
   Chagrin Falls, OH 44023

   A set consists of (1) driver pulley P/N 8-1-C and (1) driven pulley P/N 8-1-M. There is a control available with this set but you will make your own for this application.

   You can build your own variable speed pulleys like the ones from Speed Selector, Inc. used in this project but it takes lots of time. Photographs 9, 10 and 11 show a pulley I made for another project. Photo 11 also shows the fixture I made to machine the radial drive grooves in the faces of the pulleys. The 15 degree fixture is mounted on a rotary table and the pulley blank is centered atop the fixture. Machining these pulleys was a fun project when I did it but I didn't take the time to machine the ones for this job. As I said, there is a lot of time involved in machining these pulleys but you know the value of your time and I don't.

6. (5) B size V-belts ---- B51, B46, (2) B45's, and a B60. Remember that these belts will work for you only if you maintain the shaft center distances shown here. I suggest that you not purchase your belts until you have completed the installation of the drive. Then measure and order as required.

7. (2) 3"-diameter sprockets and about 4' of bicycle chain. Some sort of handwheel will be necessary here. I used one taken from an old hospital operating table (I had my appendix removed on this table when I was 11 years old, by the way) for my original saw but I fabricated the one in this project from aluminum.

8. A small bucketful of keys, bolts, nuts, setscrews and other assorted fasteners. I live close to a hardware store and it will help if you do, too.

9. Miscellaneous steel tubing, plate, angle iron, etc. for motor and bearing mounts.

electrode is adequate here. After welding, chuck the part in the lathe with the hub towards the tailstock. Machine the face of the pulley, the face of the hub and the outside diameter of the hub. Do not machine the bore at this time. Clean up the surface of the weld here if you like but be careful not to machine so much away from the weld that the strength is compromised. See Photo 14.

12. Left, weld prep. Partially welded pulley.
13. Below, Welding pulley on positioning table.

Turn the part around and chuck on the hub and proceed to finish machining the pulley. *Machinery's Handbook* suggests that for B-size V-belt pulleys under 7" in diameter the included angle should be 34/35 degrees. For 7"-and-over pulleys, use an included angle of 37/38 degrees. The only way I have of producing those angles is through the use of the compound rest. Be careful and patient in hand feeding the tool while machining these grooves because it is important that the finish on the flanks of the belt grooves be as smooth and free of tool marks as possible. The other important thing to remember is to machine the grooves deep enough so that the belt does not bottom out in the bottom of the groove. V-belts, by the way, when properly selected and installed, provide a higher degree of power transmission efficiency than do gears. Gears, unless they are designed with anti-backlash features, have that tiny amount of rotational movement before the slack is taken up and V-belts do not. I read that piece of information in *Machinery's Handbook* so you know there is no way to challenge the truth of such a statement. The drawings of the pulleys give the details of the grooves on each pulley.

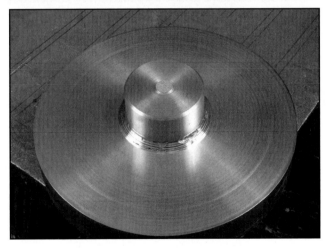

14. Pulley welded and partially machined.

Pulleys 5 and 6 are the pulleys located on the Hi-Lo shaft and are shown in Drawings No. 7 and No. 10. These pulleys are designed so that when one is driving the other will be idling on the same shaft but at a different speed from that of the shaft. All of the other pulleys will have keyways and setscrews in them. There are a bunch of keyways to cut, both internal and external. The 1/4" and 3/8" keyways were broached using a hydraulic press. These can be cut using an arbor press but it takes lots of elbow grease! The 3/16" keyways were cut in an arbor press because it was a little faster. Photos 15 and 16 show the internal keyways being cut.

There are several industrial suppliers who supply shafting with precut keyways in lengths of up to 72" and you may be able to save a little time if you choose this option. But it isn't much of a job to set up the milling machine and cut all of the keyways in one session. Good design requires that there be two setscrews in each pulley, one installed so as to bear directly on the key and one at 90 degrees to the keyway as shown in the drawings. For the larger pulleys you will need an extra length drill bit and a pulley tap of the appropriate

15. Broaching 1/4" keyway in hydraulic press.

16. Broaching 3/16" keyway in arbor press.

17. Left, sliding dog components.
18. Center, sliding dog assembled.
19. Bottom, motor mount.

size. The setscrews used here are 5/16"-18. Drawing 12 provides details of the Hi-Lo shaft and Drawing 13 is of the straight shafts.

I have labeled the shafts for purposes of discussion numbers 1 through 5 with shaft number1 being the motor and shaft number 5 being the bottom blade wheel. Shafts 1, 2 and 3 are on the bottom level and shafts 4 and 5 are elevated. I have tried in this arrangement to minimize the space it takes up yet leave enough room for assembly and maintenance. There are adjustments provided for tensioning all of the belts. The reason for the idler between pulley 10 and the bottom blade wheel is that the blade wheel cannot be moved laterally for belt tensioning. It must stay in line with the top wheel of the saw. Other belt adjustments are provided by the use of slots where the bearings are mounted.

Machine the Hi-Lo sliding dog parts as shown in Drawings 14 and 15. Drawing 14 also includes the details for the idler shaft which mounts on the frame between shafts number 4 and shaft number 5. Photos 17 and 18 show all of the parts required for the sliding dog separately and assembled. I thought when I built the original machine that the bronze split ring which

is attached to the outer ring of the dog would be a part which would require replacement from time to time, but it has been in use now for over 30 years and when I disassembled it for the purpose of making these drawings (actually I had forgotten how it was made in the intervening time!) I found that there was hardly any wear at all. Install the grease fitting as a means for lubricating the sliding dog and be sure to use it after every couple of hundred hours of use. The sealed bearings in the pulleys do not require lubrication but since the pulleys will rotate at a different speed from that of the shaft, depending upon which speed range you are in, there must be adequate clearance between the bore of the pulley and the shaft. The locknuts and snap rings serve to keep everything in place along the shaft.

The shifting fork may be either burned or sawed from steel or sawed from aluminum. The one shown here was sawed from 1" aluminum. If you burn this one out you will be able to saw the next one out with your new bandsaw. The shifting mechanism requires a bell crank so that the shift lever may be mounted at a convenient place on the front of the machine. The detent ball serves to keep it from drifting out of the selected speed range. The series of photographs show the mounting of the shifting lever, the bell crank and detent as well as the hand wheel for adjusting the Speed Selector pulleys. The sprockets and chain are necessary only to move the axis of the speed adjusting control wheel to a more convenient place for the operator. A couple of bicycle sprockets and chain will do here. Details of the detent mechanism are shown in Drawing 15 while the bell crank and the shift fork are on Drawing 16.

While you are still at the lathe, machine the rest of the parts required for the shifting controls. This includes the hand wheel and its handle on Drawing 17, the rod ends and the adjuster nut housing on Drawing 16, and the adjusting screw and nut, also shown on Drawing 17. You will need all of these parts when you start assembly of the drive and its speed adjusting controls.

The installation of the drive begins with the motor mount. A simple motor mount is shown in Photo 19. It is a sturdy arrangement and provides for over 2" of belt adjustment. I don't provide more detail than that here because your motor will be different. Just make sure that you leave adequate lateral adjustment for belt tensioning and that the motor can be bolted solidly to the base.

Build and install the bearing pedestals according to Drawings 18, 19 and the isometric Drawing 20. The self-aligning pillow block bearings are pretty forgiving if there is some minor misalignment, but you should still make the mounts so that the shafts will run as nearly parallel to the base and to each other as is possible.

If you use the Speed Selector, Inc., fixed center variable speed pulleys they will provide you with a chart for the distance between centers for a variety of belt lengths. Maintain this distance to be able to take full advantage of the system. Otherwise you run the risk of either not having the full range of speeds available or reduced belt life. The dimension

As I get older it gets harder to see through my welding hood. I have found that you can make a pretty big mess between the time you strike the arc and the time you find the crack.

given here, 15.4", is for a B46 belt. This distance works well because it provides enough space between shaft number 2 and shaft number 3 to build the mounting platforms for the bearings for shaft number 4.

Shaft number 5, the shaft which drives the bottom wheel of the saw, must be mounted so that the periphery of the blade wheel is exactly tangent to the vertical line of the blade. Since those points will not be established until you are ready to mount the table and the top wheel you have a tolerance here of plus or minus about 1/2".

The Hi-Lo shafts, shafts number 3 and 4, are one of the things which make this saw unique in its ability to provide speed ranges which will serve for metal cutting as well as for the higher speeds needed in sawing wood or synthetic materials. The way that it works is that when the sliding dog is engaged with the 3.6" diameter pulley it drives a 9.9" pulley for a speed reduction of 2.75:1. When the dog is slid to engage the 9.9" pulley the speed reduction then becomes a speed increase with a corresponding ratio. The variable speed pulleys serve to adjust between the lowest speed of the high range and the highest speed of the low range. I know — it is a head-scratcher! But it works, and has worked for me for going on 30 years.

After you get all of the bearing pedestals fabricated and welded into place you are ready to make and install the Hi-Lo shifting mechanism. Refer to the drawings to make the detailed parts and install them so that they are roughly in line vertically with the centerline of the Hi-Lo shaft itself. The shifting fork has slots where it is attached to the sliding dog so that if you need to move the shaft for belt tensioning you may do so without having an effect on the alignment.

Make and install the detent housing, the bell crank and its mounting brackets and the linkage for the shifting lever. There are adjustments built into all of these components and they will require some fine tuning to work with each other after the machine is running.

Outside of that there is nothing to it! The series of photographs at the end of this section are of several views of the parts and their fabrication and may serve to help with the inevitable questions I have left unanswered.

The wiring of your machine will depend upon your choice of motor and controls. I have three phase power in my shop and the motor I used here is a 1 HP, three phase motor. I had a combination fused disconnect switch and motor controller which worked well but I needed to have the push button start/stop station on the front of the machine so I had to do a little re-wiring. I also wired for a 110V utility receptacle and a work light. Your wiring will depend upon your choice of motor but I do suggest that you place the switch where it is conveniently located. Route the wire through the drive compartment and use straps to fasten it securely to the base or to the frame. You don't want a rotating pulley or a belt to chafe a hole through the wire insulation. Smoke and fire coming from an enclosed mechanical drive mechanism is not a good thing.

I once worked as a weldor where there was a radiographer who never made any mistakes. I didn't trust him any more than I do anybody else who is never wrong.

Series A:

A1. Right, machining angled face of pulley groove.
A2. Below left, checking for belt fit.
A3. Below right, installing setscrews. Note extra length drill and pulley tap.

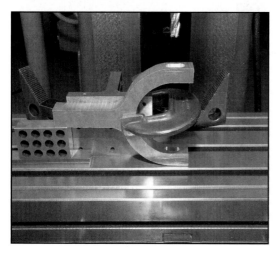

A4. Above, machining the sliding dog shifter fork.
A5. Right, Hi-Lo shaft components.

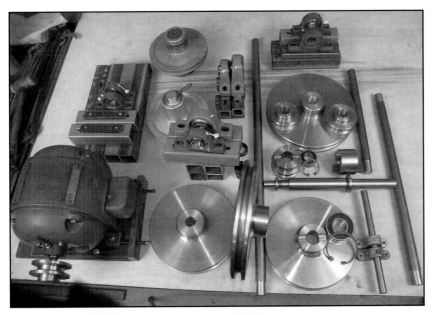

A6. Left, drive components including material for bearing pedestals.
A7. Below left, drive components in place.
A8. Below, Base with bearing pedestals and motor mount in place.

A9. Left, motor wired with conduit secured.

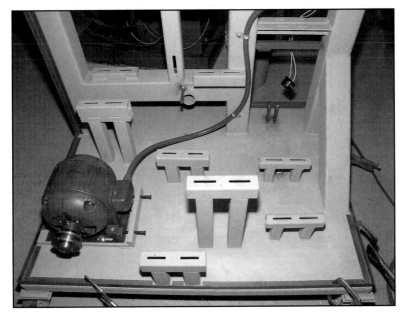

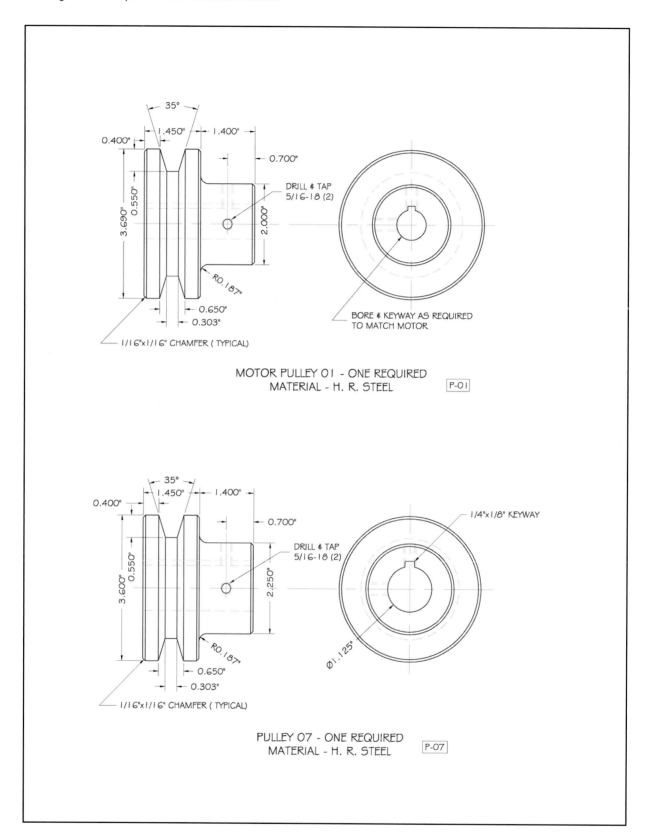

MOTOR PULLEY 01 - ONE REQUIRED
MATERIAL - H. R. STEEL   P-01

PULLEY 07 - ONE REQUIRED
MATERIAL - H. R. STEEL   P-07

Drawing No. 6.

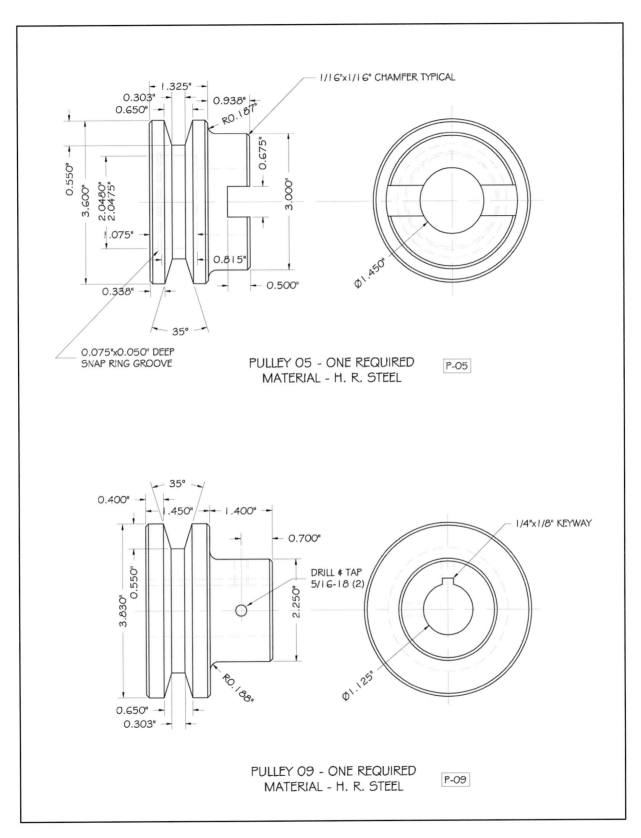

1/16"x1/16" CHAMFER TYPICAL

1.325"
0.303"
0.650"
0.938"
R0.187"
0.675"
0.550"
3.600"
2.0480"
2.0475"
3.000"
.075"
0.815"
0.338"
0.500"
35°
Ø1.450"

0.075"x0.050" DEEP
SNAP RING GROOVE

PULLEY 05 - ONE REQUIRED
MATERIAL - H. R. STEEL
P-05

35°
0.400"
1.450"
1.400"
0.700"
0.550"
3.830"
DRILL & TAP
5/16-18 (2)
2.250"
1/4"x1/8" KEYWAY
R0.188"
0.650"
0.303"
Ø1.125"

PULLEY 09 - ONE REQUIRED
MATERIAL - H. R. STEEL
P-09

Drawing No. 7.

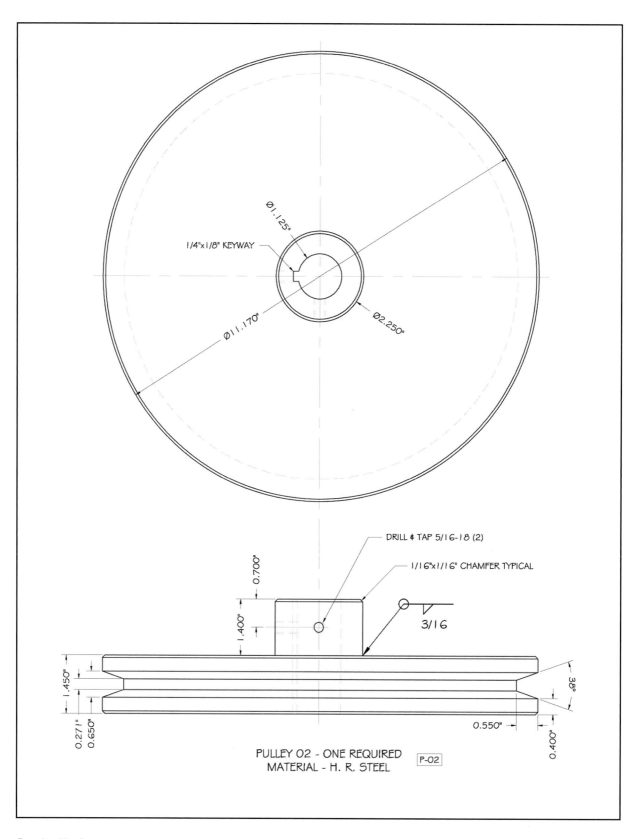

Ø1.125"

1/4"x1/8" KEYWAY

Ø11.170"

Ø2.250"

DRILL & TAP 5/16-18 (2)

1/16"x1/16" CHAMFER TYPICAL

0.700"

1.400"

3/16

1.450"

38°

0.271"
0.650"

0.550"

0.400"

PULLEY 02 - ONE REQUIRED
MATERIAL - H. R. STEEL

P-02

Drawing No. 8.

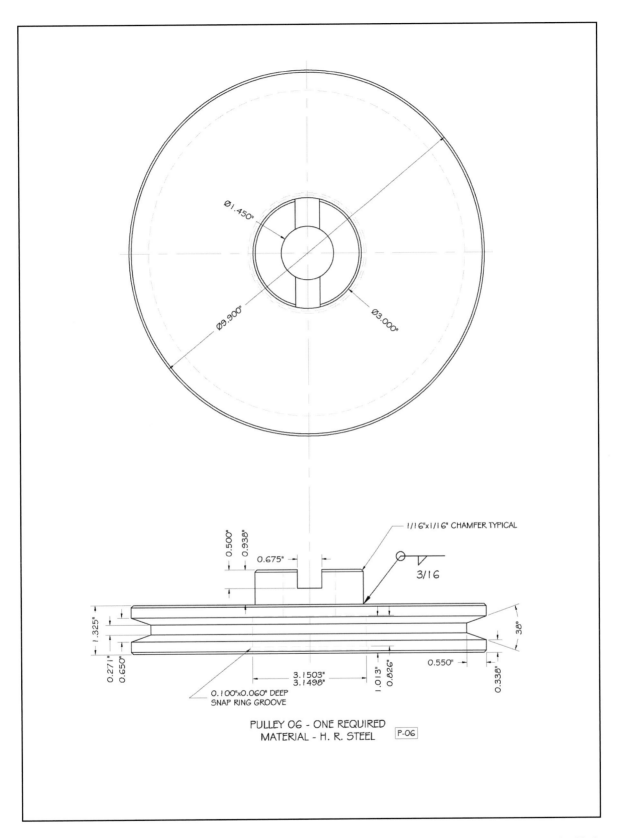

Ø1.450"

Ø9.900"

Ø3.000"

1/16"x1/16" CHAMFER TYPICAL

3/16

0.500"
0.938"
0.675"

1.325"
0.271"
0.650"

3.1503"
3.1498"

1.013"
0.826"

0.550"

38°

0.338"

0.100"x0.060" DEEP
SNAP RING GROOVE

PULLEY 06 - ONE REQUIRED
MATERIAL - H. R. STEEL   P-06

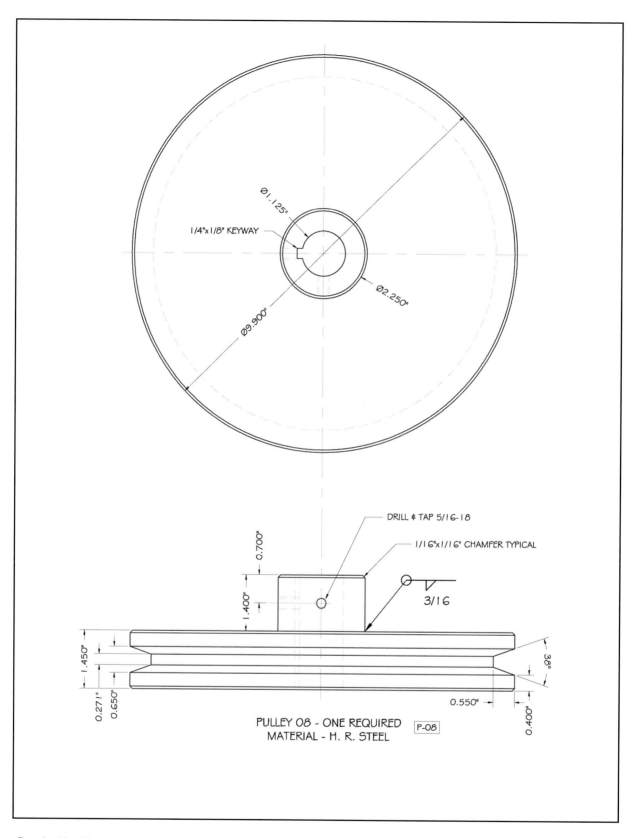

Ø1.125"

1/4"x1/8" KEYWAY

Ø2.250"

Ø9.900"

DRILL & TAP 5/16-18

1/16"x1/16" CHAMFER TYPICAL

3/16

0.700"

1.400"

1.450"

0.271"

0.650"

0.550"

0.400"

38°

PULLEY 08 - ONE REQUIRED   P-08
MATERIAL - H. R. STEEL

Drawing No. 10.

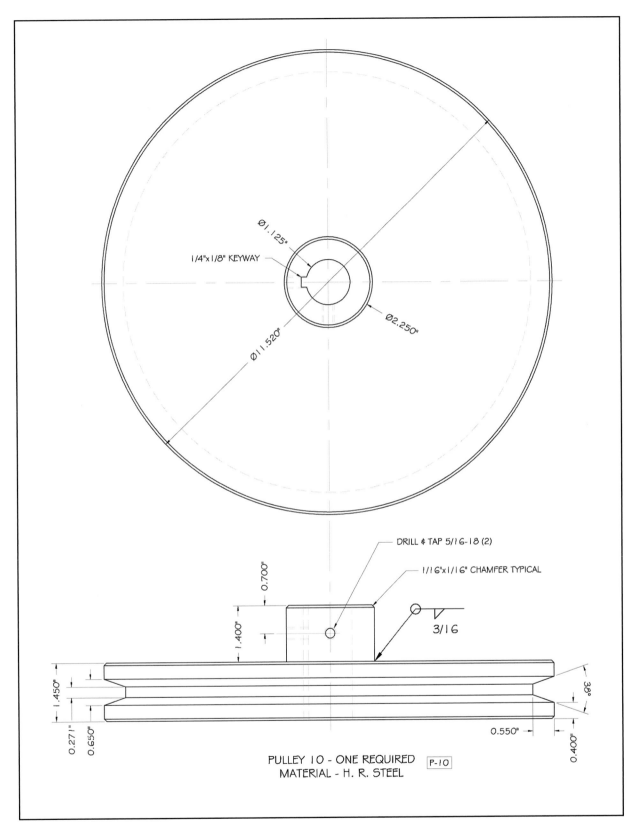

Ø1.125"

1/4"x1/8" KEYWAY

Ø2.250"

Ø11.520"

DRILL & TAP 5/16-18 (2)

1/16"x1/16" CHAMFER TYPICAL

0.700"

1.400"

3/16

1.450"

0.271"

0.650"

38°

0.550"

0.400"

PULLEY 10 - ONE REQUIRED   P-10
MATERIAL - H. R. STEEL

Drawing No. 11.

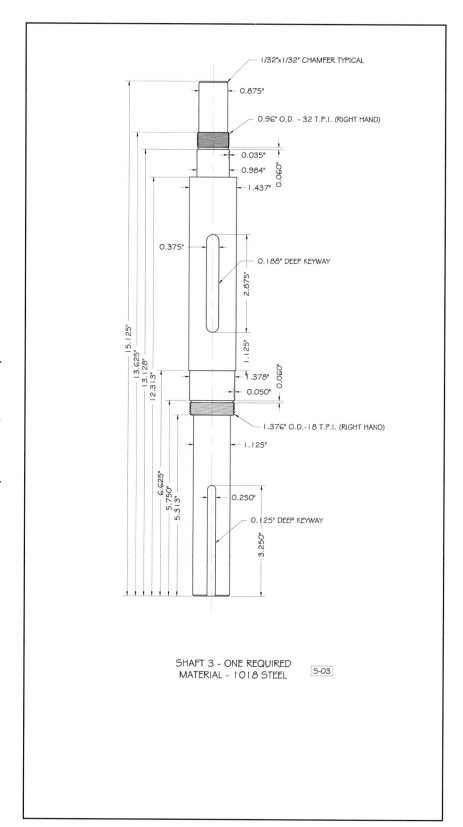

1/32"x1/32" CHAMFER TYPICAL

0.875"

0.96" O.D. - 32 T.P.I. (RIGHT HAND)

0.035"

0.984"

0.060"

1.437"

0.375"

0.188" DEEP KEYWAY

2.875"

1.125"

15.125"

13.625"

13.128"

12.313"

1.378"

0.050"

0.060"

1.376" O.D.-18 T.P.I. (RIGHT HAND)

1.125"

6.625"

5.750"

5.313"

0.250"

0.125" DEEP KEYWAY

3.250"

SHAFT 3 - ONE REQUIRED
MATERIAL - 1018 STEEL   5-03

Drawing No. 12.

The one thing that I learn more about all the time is how many things there are that I need to learn more about.

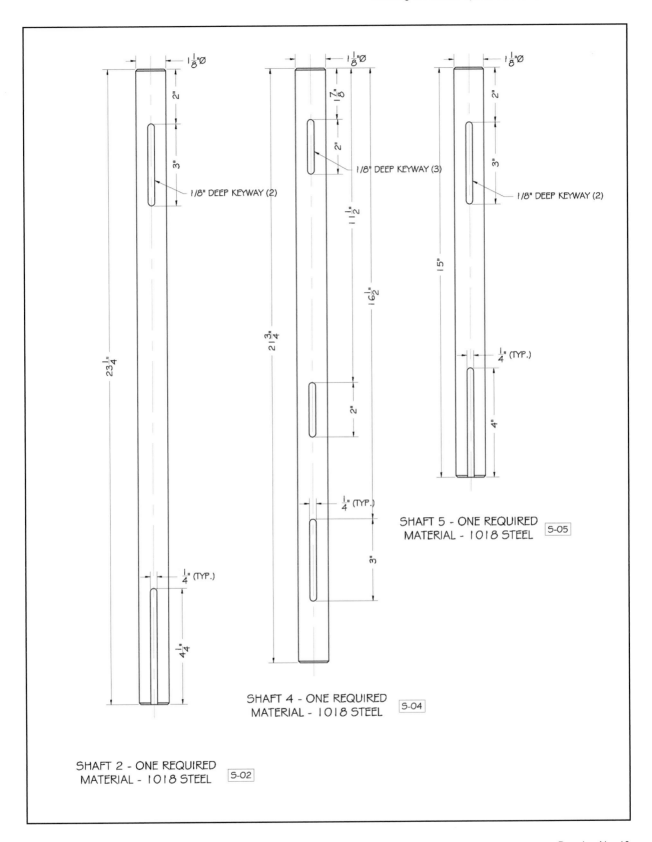

$1\frac{1}{8}"\emptyset$

2"

3"

1/8" DEEP KEYWAY (2)

$23\frac{1}{4}"$

$\frac{1}{4}"$ (TYP.)

$4\frac{1}{4}"$

SHAFT 2 - ONE REQUIRED
MATERIAL - 1018 STEEL   S-02

$1\frac{1}{8}"\emptyset$

$1\frac{7}{8}"$

2"

1/8" DEEP KEYWAY (3)

$11\frac{1}{2}"$

$16\frac{1}{2}"$

$21\frac{3}{4}"$

2"

$\frac{1}{4}"$ (TYP.)

3"

SHAFT 4 - ONE REQUIRED
MATERIAL - 1018 STEEL   S-04

$1\frac{1}{8}"\emptyset$

2"

3"

1/8" DEEP KEYWAY (2)

15"

$\frac{1}{4}"$ (TYP.)

4"

SHAFT 5 - ONE REQUIRED
MATERIAL - 1018 STEEL   S-05

Drawing No. 13.

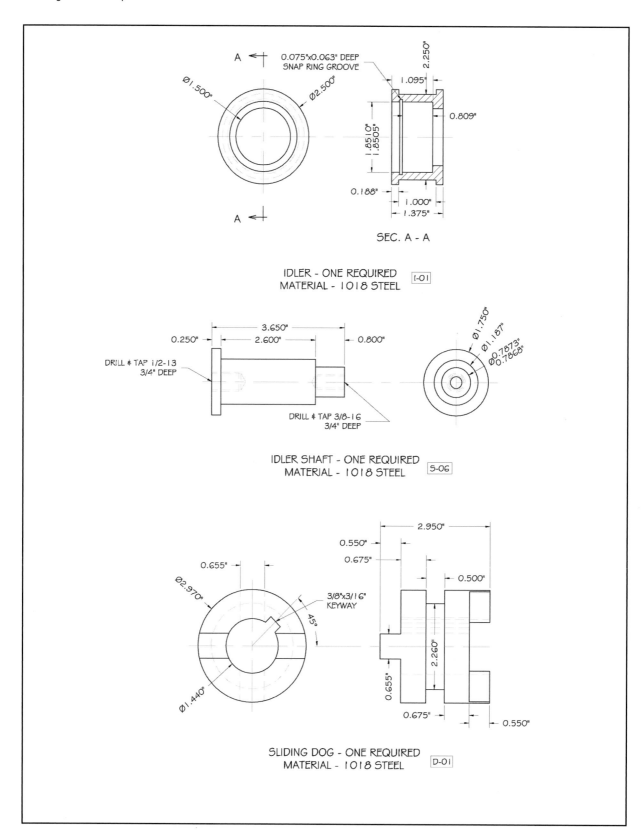

A ←

0.075"x0.063" DEEP
SNAP RING GROOVE

Ø1.500"   Ø2.500"

2.250"
1.095"
0.809"
1.8510"
1.8505"
0.188"
1.000"
1.375"

A ←

SEC. A - A

IDLER - ONE REQUIRED   [I-01]
MATERIAL - 1018 STEEL

3.650"
0.250"   2.600"   0.800"

DRILL & TAP 1/2-13
3/4" DEEP

DRILL & TAP 3/8-16
3/4" DEEP

Ø1.750"
Ø1.187"
Ø0.7873"
Ø0.7868"

IDLER SHAFT - ONE REQUIRED   [S-06]
MATERIAL - 1018 STEEL

0.655"
Ø2.970"

3/8"x3/16"
KEYWAY

45°

Ø1.440"

2.950"
0.550"
0.675"
0.500"
0.655"
2.260"
0.675"
0.550"

SLIDING DOG - ONE REQUIRED   [D-01]
MATERIAL - 1018 STEEL

Drawing No. 14.

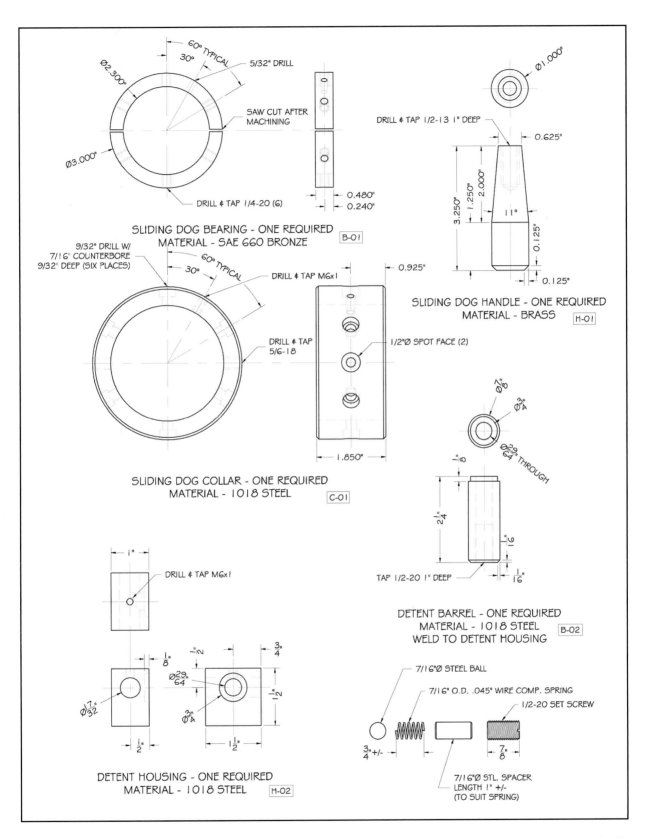

Ø2.300"
60° TYPICAL
30°
5/32" DRILL
SAW CUT AFTER MACHINING
Ø3.000"
DRILL & TAP 1/4-20 (6)
0.480"
0.240"

SLIDING DOG BEARING - ONE REQUIRED
MATERIAL - SAE 660 BRONZE   B-01

Ø1.000"
DRILL & TAP 1/2-13 1" DEEP
0.625"
3.250"
2.000"
1.250"
11°
0.125"
0.125"

SLIDING DOG HANDLE - ONE REQUIRED
MATERIAL - BRASS   H-01

9/32" DRILL W/ 7/16" COUNTERBORE 9/32" DEEP (SIX PLACES)
60° TYPICAL
30°
DRILL & TAP M6x1
DRILL & TAP 5/6-18
0.925"
1/2"Ø SPOT FACE (2)
1.850"

SLIDING DOG COLLAR - ONE REQUIRED
MATERIAL - 1018 STEEL   C-01

Ø7/8"
Ø3/4"
Ø29/64" THROUGH
1/8"
2 1/4"
1/16"
1/16"
TAP 1/2-20 1" DEEP

DETENT BARREL - ONE REQUIRED
MATERIAL - 1018 STEEL   B-02
WELD TO DETENT HOUSING

1"
DRILL & TAP M6x1

1/8"
1/2"
3/4"
Ø29/64"
Ø17/32"
Ø3/4"
1/2"
1/2"
1 1/2"

DETENT HOUSING - ONE REQUIRED
MATERIAL - 1018 STEEL   H-02

7/16"Ø STEEL BALL
7/16" O.D. .045" WIRE COMP. SPRING
1/2-20 SET SCREW
3/4" +/-
7/8"
7/16"Ø STL. SPACER LENGTH 1" +/- (TO SUIT SPRING)

Drawing No. 15.

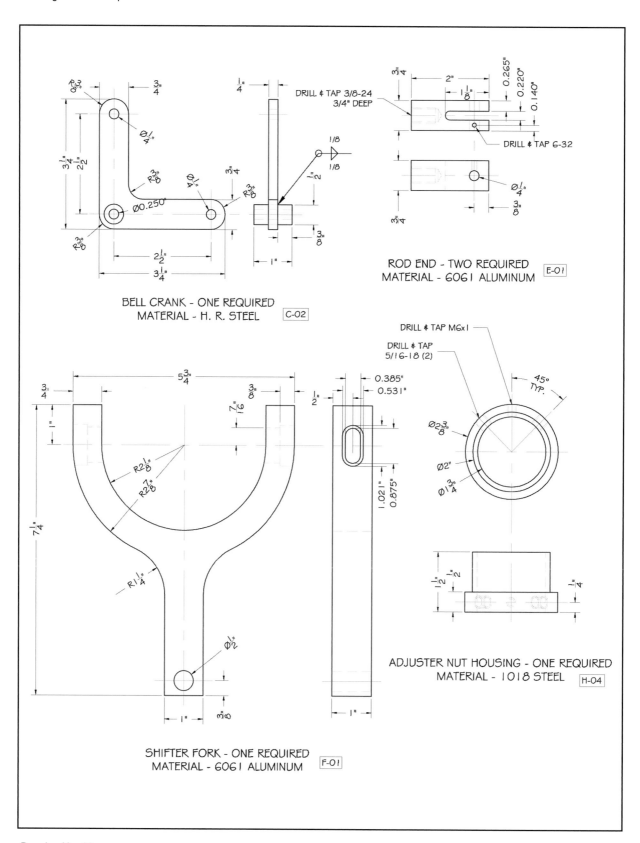

BELL CRANK - ONE REQUIRED
MATERIAL - H. R. STEEL   C-02

DRILL & TAP 3/8-24
3/4" DEEP

ROD END - TWO REQUIRED
MATERIAL - 6061 ALUMINUM   E-01

DRILL & TAP M6x1

DRILL & TAP
5/16-18 (2)

ADJUSTER NUT HOUSING - ONE REQUIRED
MATERIAL - 1018 STEEL   H-04

SHIFTER FORK - ONE REQUIRED
MATERIAL - 6061 ALUMINUM   F-01

Drawing No. 16.

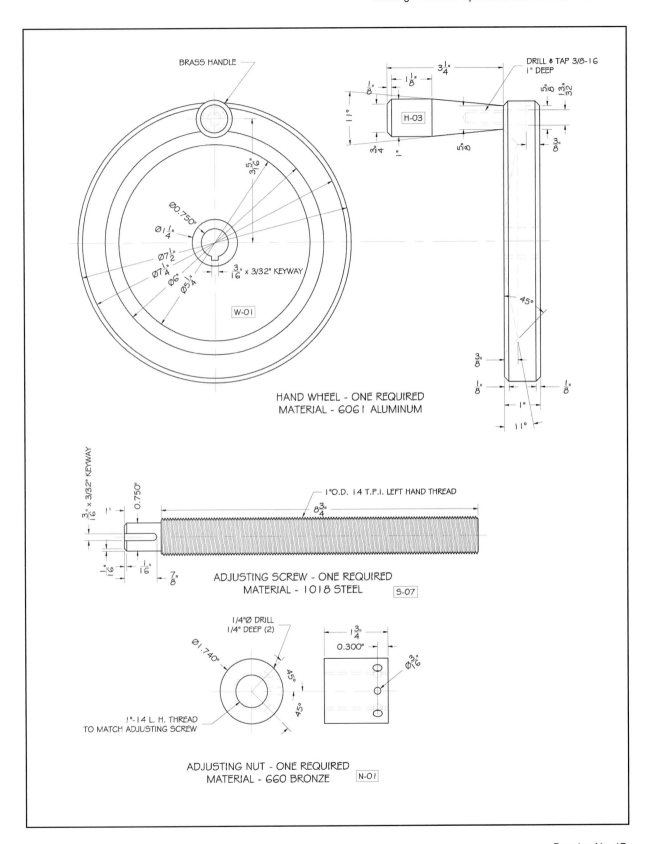

BRASS HANDLE

DRILL & TAP 3/8-16
1" DEEP

H-03

Ø0.750"
Ø1¼"
Ø7½"
Ø7¼"
Ø6¼"
Ø5¼"

3/16" x 3/32" KEYWAY

W-01

HAND WHEEL - ONE REQUIRED
MATERIAL - 6061 ALUMINUM

45°

11°

3/16" x 3/32" KEYWAY

0.750"

1"O.D. 14 T.P.I. LEFT HAND THREAD

8¾"

ADJUSTING SCREW - ONE REQUIRED
MATERIAL - 1018 STEEL   S-07

1/4"Ø DRILL
1/4" DEEP (2)

Ø1.740"

45°

45°

1"-14 L. H. THREAD
TO MATCH ADJUSTING SCREW

1¾"

0.300"

Ø3/16"

ADJUSTING NUT - ONE REQUIRED
MATERIAL - 660 BRONZE   N-01

Drawing No. 17.

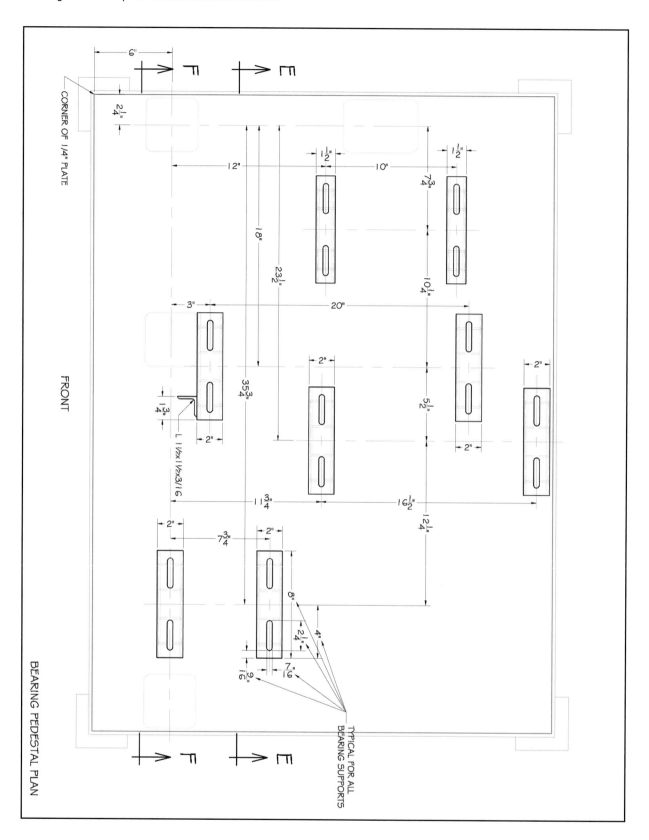

Drawing No. 18.

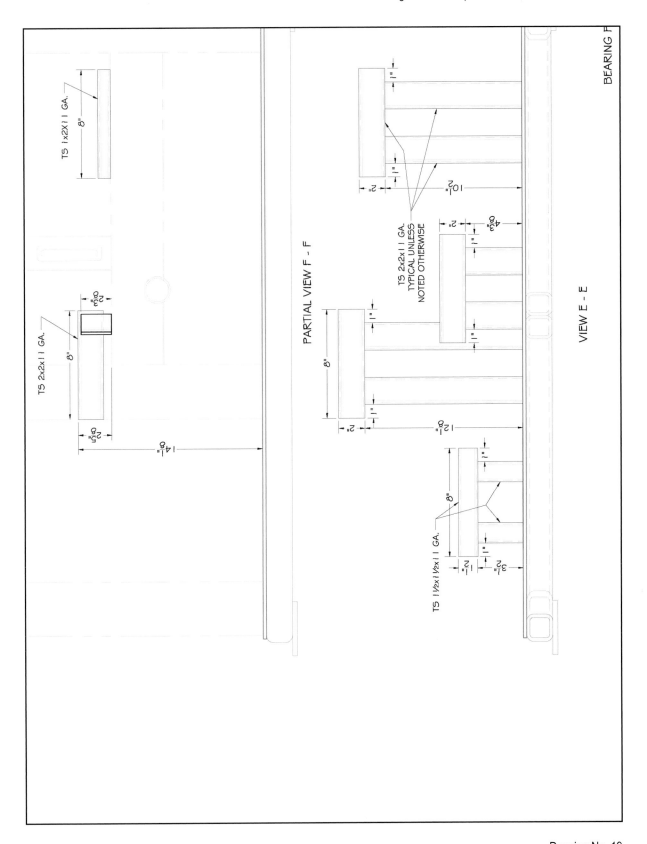

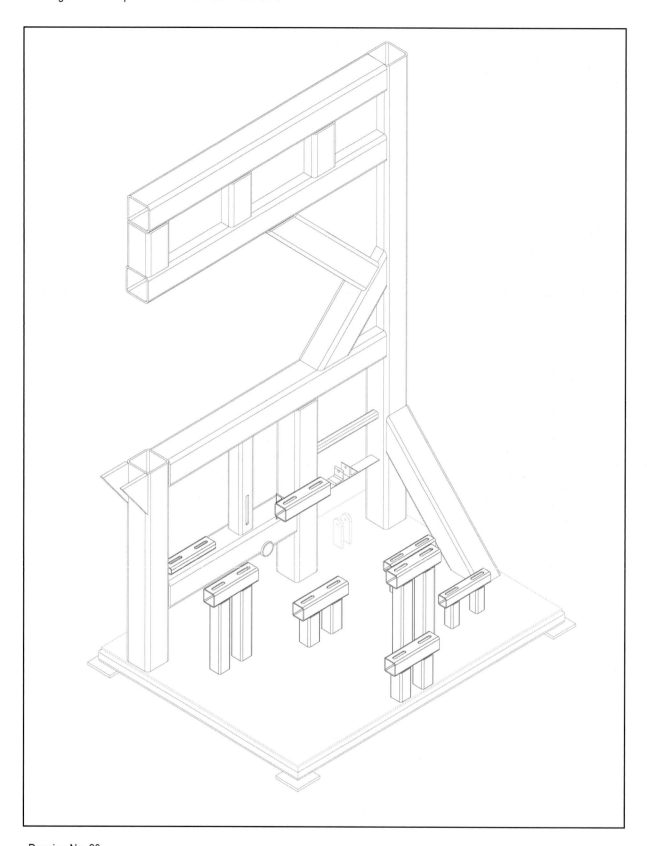

Drawing No. 20

SECTION THREE: Blade Wheels, Table and Guides

I am using a little different approach to this section. I am going to provide for you detailed drawings of the wheels, the wheel spindles, the table and its associated mounting parts, and of some of the blade guide holding hardware. But I am not going to try to detail the guides themselves because there are so many options available. I will provide instead several photographs with detailed captions. Hopefully, with those photos and a lot of talking about them on my part there will be enough information for you to make the choices and build the guide assemblies.

Before we begin the fabrication of the table itself we must finish up the table mounting plate which was first mentioned in Section One. The details of the plate and its location relative to the frame of the saw are shown on Drawing 21. The position of this plate is important in that it will determine the final position of the table and, consequently, the

Getting a cast iron part too clean to braze is a little like finding an NBA candidate who is too tall.

Material List
Blade Wheels, Table and Guides

1. (3) pieces of 6061-T651 aluminum plate — 1-1/2" thick sawed to 12-3/4" in diameter — both idler wheels and the bottom or drive wheel. For the idler wheels you will also need (2) pieces of 3/4" aluminum plate 5" in diameter.

2. (1) piece of 5/8" steel plate — 4-3/4" square — mounting plate for offset wheel.

3. (1) piece of 5/8" X 4" CRS flat bar — 6" long — mounting plate for top wheel.

4. (2) pieces of 1-1/2" round CRS — 4" long — axle shafts for both idler wheels. Both axles will require locknuts and I machined them by cutting down the width of standard 3/4"-16 nuts and installing setscrews.

5. (2) A-6 Federal Mogul inner spindle bearings.

6. (2) A-3 Federal Mogul outer spindle bearings. The wheel bearings are standard automotive front spindle bearings available from auto parts suppliers.

7. (2) 473215 Federal Mogul grease seals — axle seals for the back side of the bearings.

8. (2) 2" diameter freeze plugs — Dorman P/N 555-044 — used as hub caps for the wheel bearings.

9. (1) piece of 3/4" A36 steel plate — burned to 26" X 26" — Saw table proper. If you have a good supplier whose abilities you can depend upon you can have this part burned out to size. A good burning machine will produce an edge that will serve without having to be machined. Otherwise you should machine at least one edge of the table.

10. 1" X 2" HRS steel bar — (2) pieces 21-1/2" long — under support for the table.

11. 3/4" X 2" HRS steel bar — (2) pieces 24" long, (2) pieces 11" long and (1) piece 22-1/2" long — under support for the table.

12. 1/2" X 2" HRS steel bar — for making the table insert.

13. (2) pieces of 3/4" X 2" flat bar X 14" long — table rails.

14. (2) pieces of 1/2" X 2-1/2" flat bar X 22-1/2" long — table rails.

15. (2) pieces of 1/2" X 3" flat bar X 22-1/2" long — table rails.

16. 1 piece of CRS bar 1" X 1-1/2" X 18" long — upper blade guide mounting bar.

17. Approximately 8' of 3/32" aircraft control cable or equivalent — cable clamps to fit.

18. An 8 pound piece of scrap which will fit into the 4" square tubing frame. This piece and the cable will provide the counterweight for the upper blade guide bar.

1. Left, the bottom side of the welded table.
2. Above, machining the table slot to fit the insert.

table slot and the blade. Locate the plate at this time and weld it into place, including the gusset plates which support the outer portion.

Now to the table itself. Saw all of the under support pieces to length and tack them into place. Refer to Drawing number 22 for the locations of the under supports or ribs. There is a lot of good welding here, right there in front of you on the welding table. You can sit down and do most of it. I do some of my most profound thinking while I am welding something like this which probably has some meaning but I have no idea what. You can drill and tap the 1/2"- 13 holes in the 1" bars before welding them in but be careful about alignment when you begin welding. And be sure to leave bolts in the holes while you are welding to keep the spatter out of the threads. Drill a 1-1/2" hole through the center of the table and then either saw cut or burn the slot. This should be done after the welding is done on the table. Photo 1 is of the completed welding and Photo 2 is of the slot being finish machined.

3. Angle iron clips welded in to hold the insert in place while being ground.

Machine the table insert, part number I-02 shown on Drawing 23, at this time and weld a couple of pieces of scrap angle iron to the insert and the table on the bottom side. See Photo 3. These will be removed when the table comes back from the grinding shop. The purpose for this is to have the insert in place when the table is ground so that the table surface will be smooth and uninterrupted. After completing the welding and the installation of the table insert

4. Machining the back side of one of the upper wheels. Notice the piece of 3/4" aluminum plate bolted on to give added thickness for the bearings.

Every can of paint has a run in it somewhere.

take the assembly to wherever you have Blanchard grinding done. This part of the process is necessary for a number of reasons. There will be a slight convexity in the table due to the stresses introduced by the welding. There is also the mill scale on the plate. Grinding the finished table with its insert in place will give the whole job a more finished appearance. If you have your own grinder then send me your address and your rates. I frequently have to take parts out for grinding and I use a shop in Fletcher, NC called TDM Associates. They have always done a good job for me and their prices are fair.

On my original table I machined a miter groove 6" to the right of, and parallel to, the blade slot. I have never used this feature and it was probably a waste of time to machine it. But if you think you will have use for one now is the time to machine it.

While we are working on the table and its associated parts is a good time to cut, drill and weld the table support rails detailed on Drawing 29. There are several opportunities for errors to accumulate in the fabrication of the table and its hardware. We have to drill and tap the holes in the bottom of the table and in the top of the table mounting plate. We must weld up the table mounting rails and the table itself. I don't care how careful you are in these operations there will be some accumulation of errors and if life were fair they would counteract each other. But life ain't always fair and if it can go wrong it will. For that reason it is a good idea to drill the holes in the table mounting rails to 5/8" in diameter. If the holes are off by much more than that then the unfairness of life may not be the only place to put the blame.

Now let's go to the wheels and their spindles and/or axles. The two upper wheels are identical and interchangeable and are detailed in Drawing 24. You may want to choose a little different route here in your bearing selection but I used automotive front spindle bearings in my original saw and they will probably never wear out. The disadvantage is that they require a little more spindle length and that is why the piece of 3/4" aluminum plate is bolted to these wheels on the back. The automotive bearings are a little less expensive and that probably influenced my decision as well. The bottom wheel, shown as the drive wheel on Drawing 25, requires only a 1-1/8" straight bore through with a 1/4" keyway. The difficulty of installing a setscrew in this wheel is overcome by using set collars on either side of the wheel. Photo 4 shows the machining of the back side of one of the upper wheels.

Here is a good place to hold a brief discussion on the outer surfaces of the wheels. As you can see in the drawings I have machined a recessed area into the periphery of the wheels. When I built my original saw I made the wheels the same size as a Rockwell bandsaw which was on the market at that time. I purchased tires for my saw from Rockwell and they worked well for a long time. When they finally wore out, I couldn't find replacements so I made some with neoprene strips and upholsterer's glue. And that is the route I am taking with this machine. McMaster/Carr sells bandsaw wheel tires, or wheel rubbers as they call them, in various sizes for about $20 to $30 apiece, depending upon the size. There probably is a better way than the way I am doing it but I just haven't found it yet and the method I use works well

A good philosophy to live by. All guns are loaded. All snakes are poisonous. And all wires are hot. Stick by that rule and you can live longer.

5. The two spindle plates welded up and ready for the lathe.

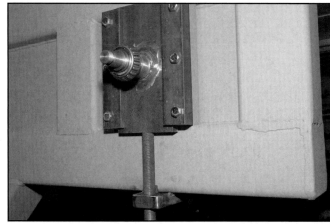

6. Above, one of two spindle nuts. Notice the 6-32 setscrew.

7. Right, blade tensioning mechanism welded into place.

Give me a 50 foot long piece of 1/0 welding lead over a parachute any day. If you find yourself in the air without an airplane one end of that sucker will hang on something!

enough that I don't have the incentive to work on it. The machine will work without using any tires at all but you run the risk of accidentally machining an unwanted groove in the outer surface of your aluminum wheels.

The spindles for the idlers differ in that the position of the top wheel spindle, part number S-08 on Drawing 26, must have the ability to be adjusted vertically in order to set blade tension. There should be at least a couple of inches of vertical adjustment here to accommodate differences in blade lengths. The 3/8"-16 holes drilled and tapped in the mounting plate for the third, or offset, wheel spindle, part number S-09 also shown on Drawing 26, are there to provide adjustments for tilting the axle a small amount in any direction. This may not be necessary but if you have trouble with the blade tracking on the wheels this will give you the means for correcting that and it is a simple matter of drilling and tapping a few holes. Machine the spindles themselves by chucking the welded assembly in a four- jaw chuck or bolting them to a face plate. Photo 5 shows the spindles welded and ready to machine. Photo 6 is of the locking nut you will need for each of the upper wheel spindles.

The vertical adjustment of the top idler is obtained by making the slides shown as part number G-01 on Drawing 23. You will need two of these parts as well as the adjusting screw, S-10, and the threaded part it passes through, B-03, shown on Drawing 27. The handwheel and its handle are detailed in Drawing 28. Photo 7 shows the assembly in position and welded onto the frame. Remember that the location of the periphery of this upper wheel will determine the position of the table slot on the table and the location of the centerline of the bottom wheel. After this wheel is put into position there are no adjustments provided for as to side to side movement of the blade location.

Here is where it all begins to come together, so pay attention. After mounting the two upper wheels, make an alignment gage, for lack of a better description, out of a piece of string and a bungee strap as shown in Photo 8. Place a good machinist's square across the table mounting plate and align the bottom wheel with the top wheel so that the blade will be perfectly square with the table. The idler you installed while building the drive mechanism will allow the belt tension to be adjusted for the drive wheel. Tighten all of the bearing mounts for the bottom wheel because this is where they will stay.

Now to the table itself. Either call your neighbors in to help with this installation or make a bridle like the one seen in Photo 9. Lift the table into place and if you have been careful with all of your dimensions and layout it will fit into place with the string from the "gage" passing exactly through the center of the 1-1/2" hole you drilled a few weeks ago. But you aren't ready to bolt it down yet! As soon as you make sure that the mounting bolts all line up and that the blade slot is in the correct position relative to the wheels, set the table back off until after you have installed the bottom blade guide.

8. Above, final positioning of the wheels to insure the blade is square with the table.
9. Below, installing the table. Notice the chain bridle.

Blade guides. This is where you will have to wing it to some extent. The series of photographs included here, Series A, will show some of the options for blade guide installation. When I built my original saw I did a lot of looking for a set of blade guides and finally found a used set at an obscure shop in Asheville, N.C., but that shop owner has long since gone out of business and retired. So that is one option I didn't have this time. There are many others. You can purchase a set of generic blade guides from McMaster-Carr. You can buy a set from DoAll or from one of the other saw manufacturers. You can modify guides from a horizontal cutoff saw or you can build your own. Whichever option you choose keep in mind that the guides must be adjustable as to both the thickness and the width of the blade. They also must give good support to the back of the blade. When you are sawing 2" thick steel plate

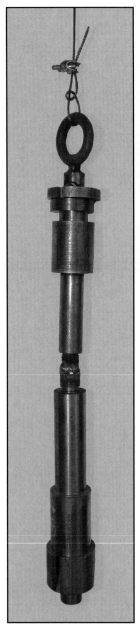

10. Above, the counterweight for the upper blade guide bar. Luckily, this will never be seen.
11. Right, counterweight pulley system.

you will be applying a lot of pressure to the blade and to the guides.

Machine the two halves of the slide, see Drawing 30, where the top guide vertical adjustment will move and weld the bottom half to the saw frame. Here is where you will use the pulleys which you installed in the fabrication of the frame long ago. After you have assembled the guide and the sliding bar it is bolted to, make a piece of art similar to the one shown in Photo 10. You can tell from the photo that the appearance of this part is not important but its weight is. If you use similar guides to the ones I used, the top guide mounting bar and the guide itself will weigh about 8 pounds. Use the piece of aircraft control cable to run from the top of the guide assembly, up to the pulley in the end of the frame arm, across to the pulley on the opposite side and down into the 4" square tubing of the frame where it will be attached to the counterweight, see Photo 11. This is not absolutely necessary for the operation of the machine but it is something you will not be sorry you took the time to do.

The guides I used on this machine are from a DoAll saw I purchased at an auction for the purpose of rebuilding and reselling. There were several sets of guides with the saw and they were right here in the shop so I used them. Notice that there are two distinct types of guide shown, one with wheel guides and one with hardened slides. The type with the slides will probably work best if you use narrow blades while the wheeled type will work with wider blades.

On this saw I made the mounting points for the guides simply as 3/4" holes drilled horizontally through the mounting points. This allows fore and aft movement of the guide when using different blade widths. Other than the points listed here, blade guides are easy!

In section Four we will finish the machine by building the housings and the doors. We are almost ready to make sawdust!

Series A

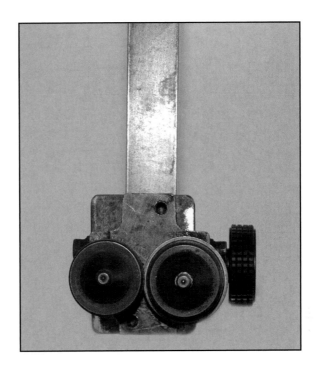

A1. Above, upper blade guide and blade tensioning device.

A2. Right, one type of blade guide.

A3. Left, upper blade guide bar. A 3/4" hole machined through with a 1/4"-20 knurled setscrew.

A4. Below left, another type of blade guide mounted in the upper position.

A5. Below, bottom blade guide mount welded into place.

A6. Top left, bottom blade guide installed.

A7. Top right, another view of the bottom guide.

A8. Above, bottom guide seen from the back.

A9. Above right, upper blade guide vertical adjustment.

A10. Right, upper blade guide adjustment and blade tensioning arrangement.

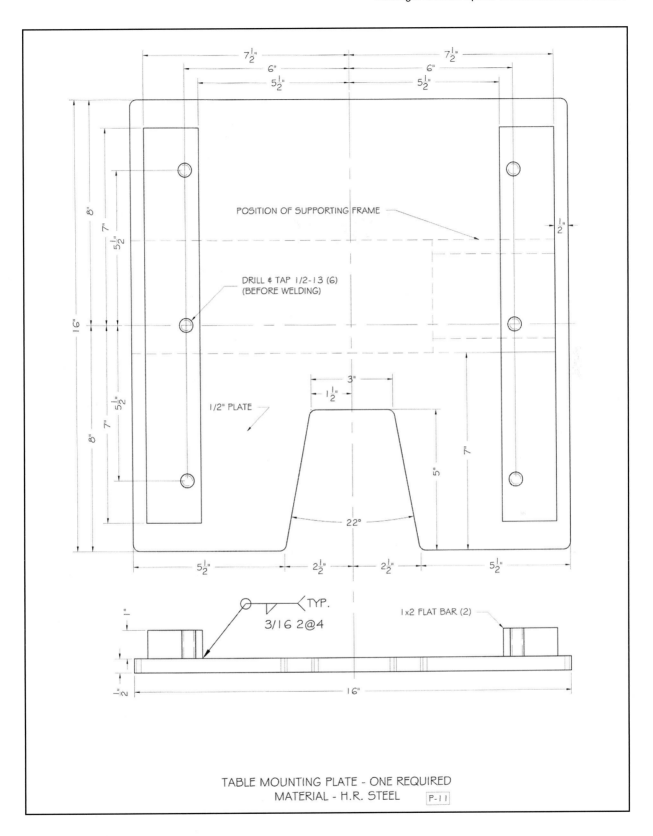

POSITION OF SUPPORTING FRAME

DRILL & TAP 1/2-13 (6)
(BEFORE WELDING)

1/2" PLATE

22°

TYP.
3/16 2@4

1x2 FLAT BAR (2)

TABLE MOUNTING PLATE - ONE REQUIRED
MATERIAL - H.R. STEEL   P-11

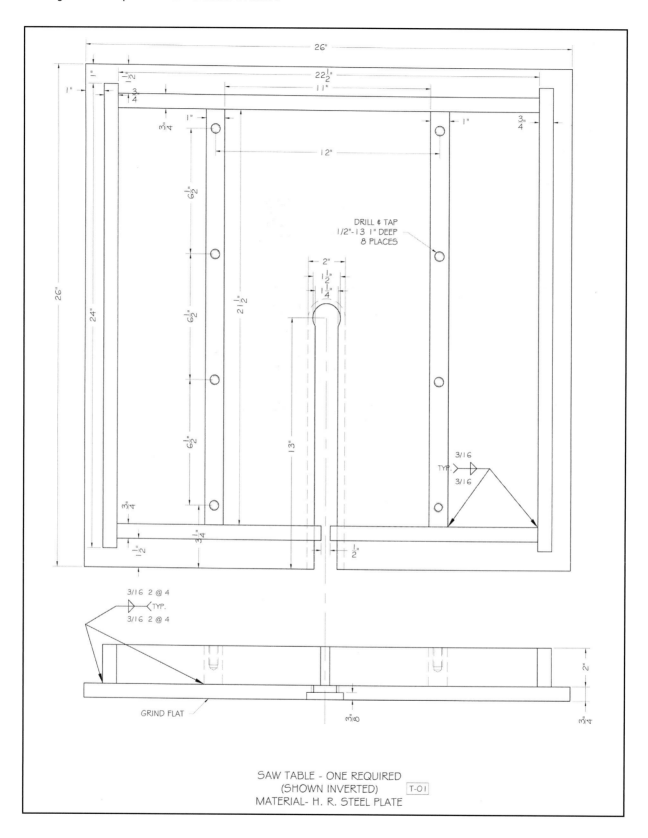

SAW TABLE - ONE REQUIRED
(SHOWN INVERTED)   T-01
MATERIAL- H. R. STEEL PLATE

Drawing No. 22.

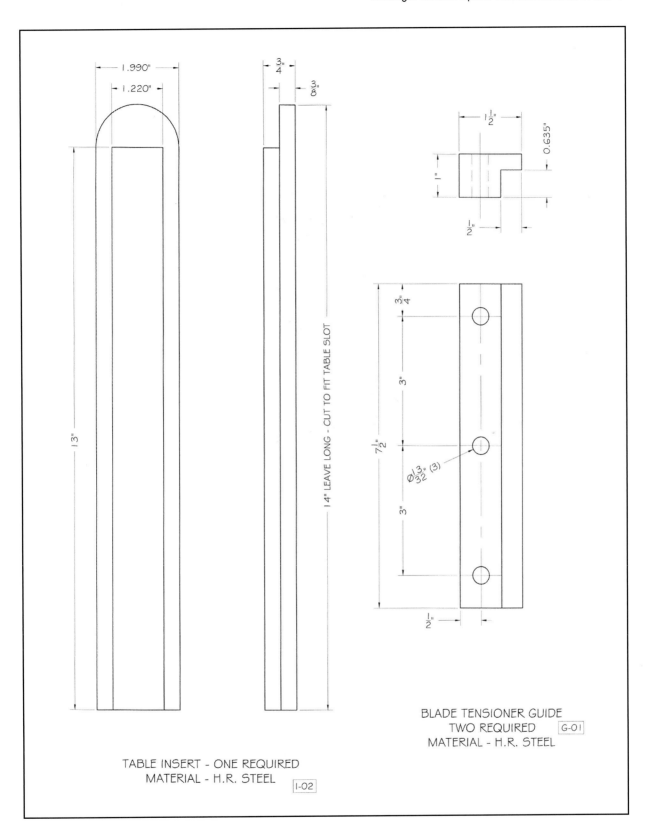

1.990"

1.220"

3/4"

3/8"

1 1/2"

0.635"

1"

1/2"

3"

13"

14" LEAVE LONG - CUT TO FIT TABLE SLOT

7 1/2"

3/4"

3"

3"

Ø13/32" (3)

1/2"

BLADE TENSIONER GUIDE
TWO REQUIRED   G-01
MATERIAL - H.R. STEEL

TABLE INSERT - ONE REQUIRED
MATERIAL - H.R. STEEL   I-02

Drawing No. 23.

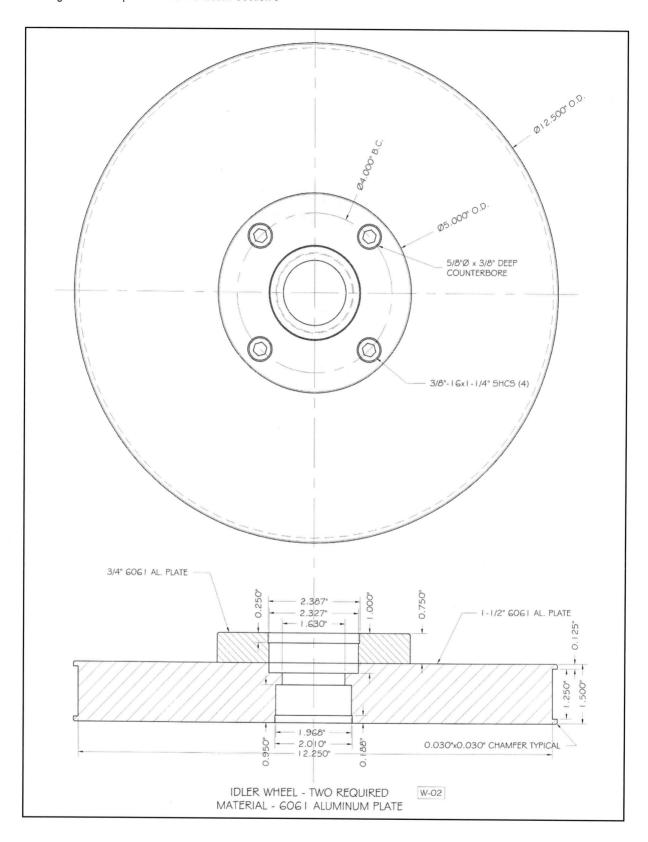

Ø12.500" O.D.

Ø4.000" B.C.

Ø5.000" O.D.

5/8"Ø x 3/8" DEEP
COUNTERBORE

3/8"-16x1-1/4" SHCS (4)

3/4" 6061 AL. PLATE

0.250"
2.387"
2.327"
1.630"
1.000"
0.750"

1-1/2" 6061 AL. PLATE

0.125"

1.250"
1.500"

1.968"
2.010"
0.188"

0.030"x0.030" CHAMFER TYPICAL

0.950"
12.250"

IDLER WHEEL - TWO REQUIRED   W-02
MATERIAL - 6061 ALUMINUM PLATE

Drawing No. 24.

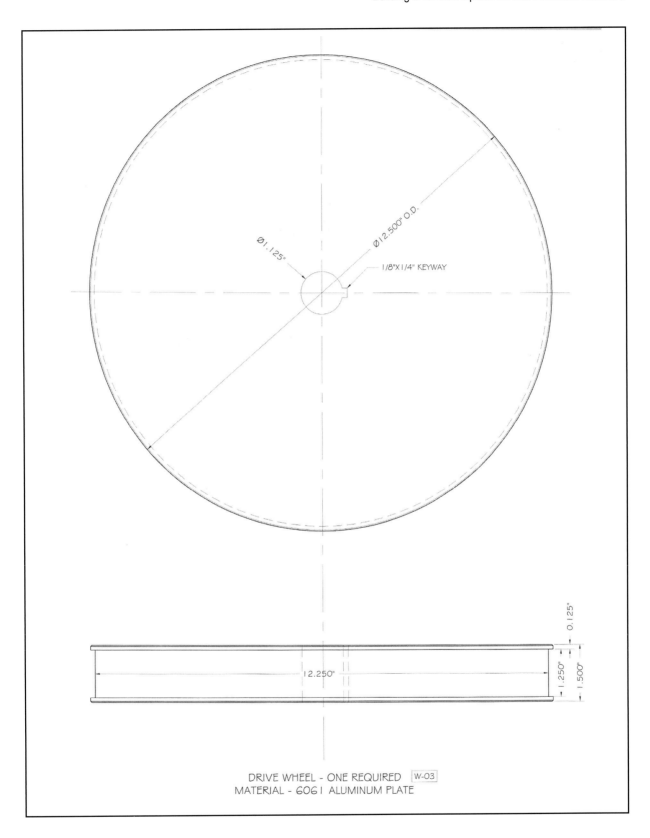

Ø1.125"

Ø12.500" O.D.

1/8"X1/4" KEYWAY

0.125"

1.250"

1.500"

12.250"

DRIVE WHEEL - ONE REQUIRED   W-03
MATERIAL - 6061 ALUMINUM PLATE

Drawing No. 25.

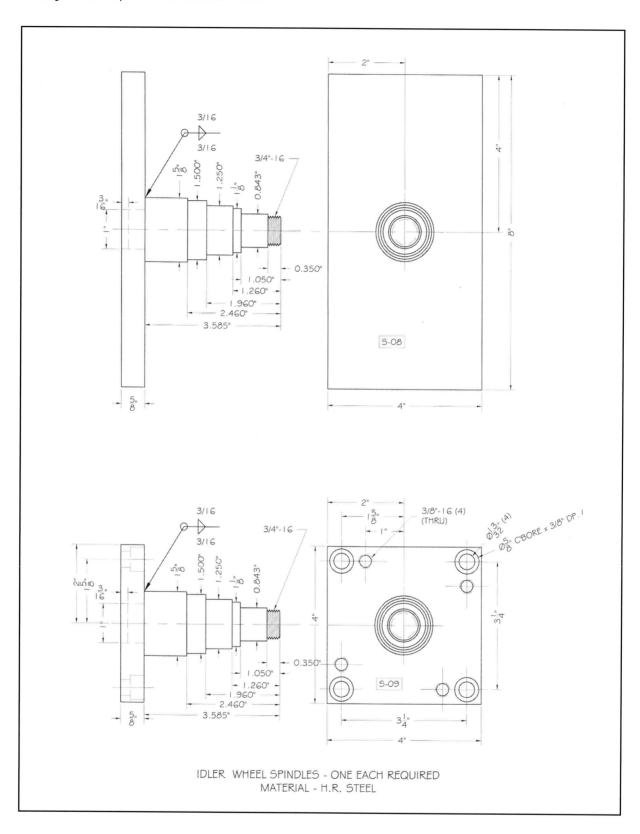

IDLER WHEEL SPINDLES - ONE EACH REQUIRED
MATERIAL - H.R. STEEL

Drawing No. 26.

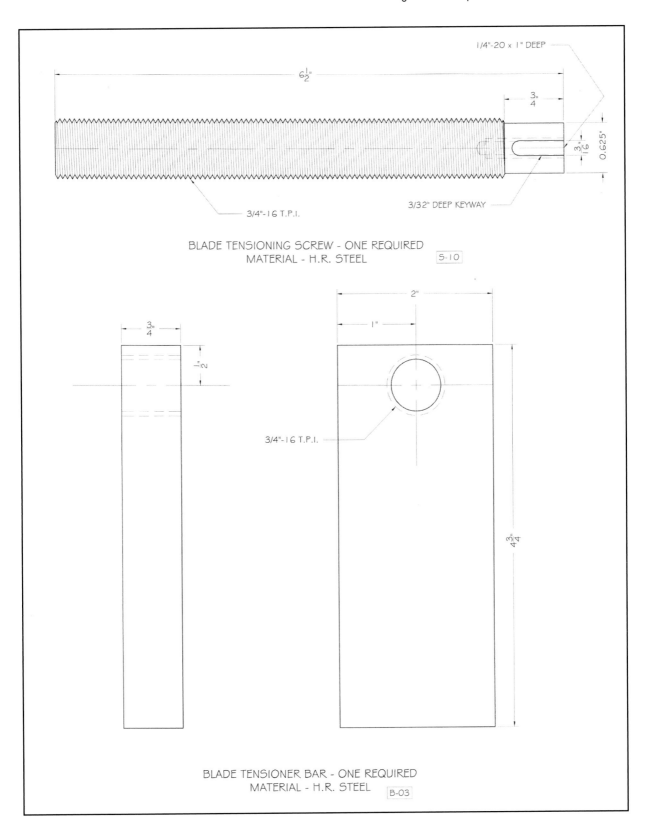

1/4"-20 x 1" DEEP

6 1/2"

3/4"

3/16"

0.625"

3/4"-16 T.P.I.

3/32" DEEP KEYWAY

BLADE TENSIONING SCREW - ONE REQUIRED
MATERIAL - H.R. STEEL   S-10

2"

1"

3/4"

1/2"

3/4"-16 T.P.I.

4 3/4"

BLADE TENSIONER BAR - ONE REQUIRED
MATERIAL - H.R. STEEL   B-03

Drawing No. 27.

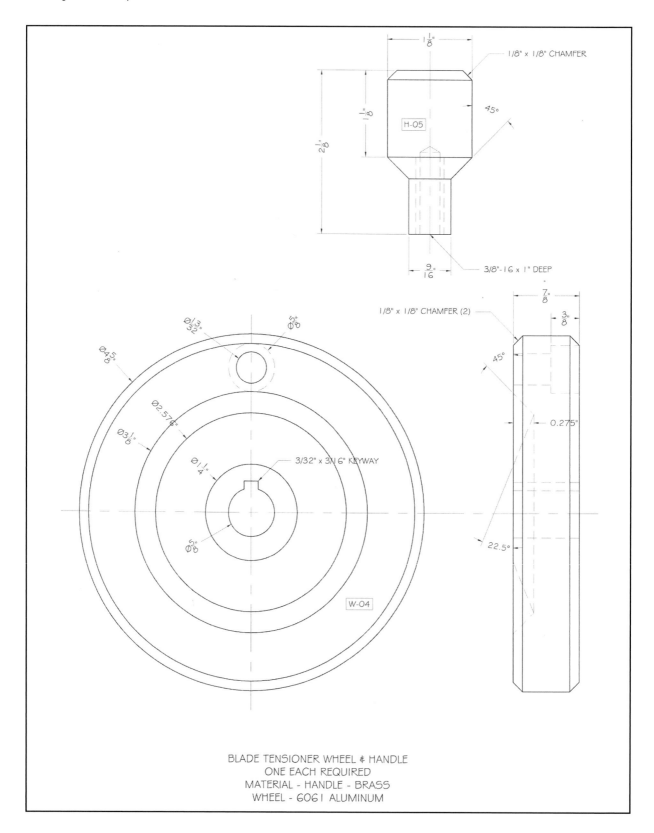

1/8" x 1/8" CHAMFER

45°

H-05

3/8"-16 x 1" DEEP

1/8" x 1/8" CHAMFER (2)

45°

0.275"

22.5°

3/32" x 3/16" KEYWAY

W-04

BLADE TENSIONER WHEEL & HANDLE
ONE EACH REQUIRED
MATERIAL - HANDLE - BRASS
WHEEL - 6061 ALUMINUM

Drawing No. 28.

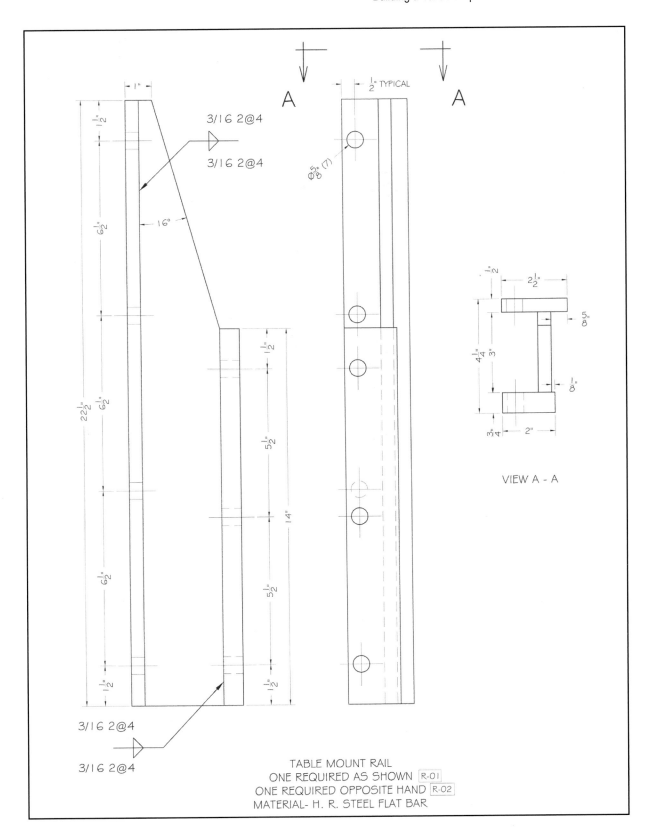

3/16 2@4
3/16 2@4

3/16 2@4
3/16 2@4

VIEW A - A

TABLE MOUNT RAIL
ONE REQUIRED AS SHOWN R-01
ONE REQUIRED OPPOSITE HAND R-02
MATERIAL- H. R. STEEL FLAT BAR

Drawing No. 29.

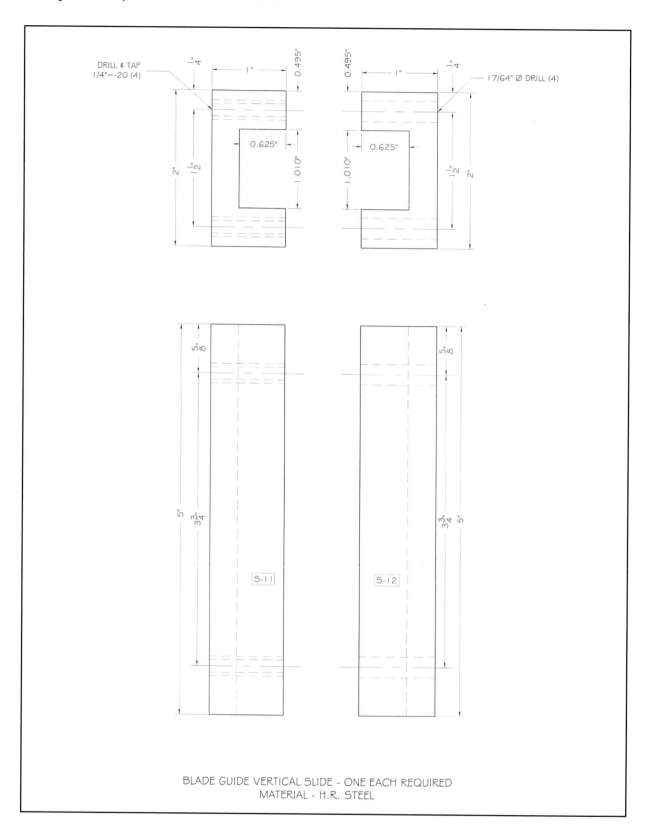

DRILL & TAP
1/4"=-20 (4)

17/64" Ø DRILL (4)

0.625"

0.625"

0.495"

0.495"

1.010"

1.010"

S-11

S-12

BLADE GUIDE VERTICAL SLIDE - ONE EACH REQUIRED
MATERIAL - H.R. STEEL

Drawing No. 30.

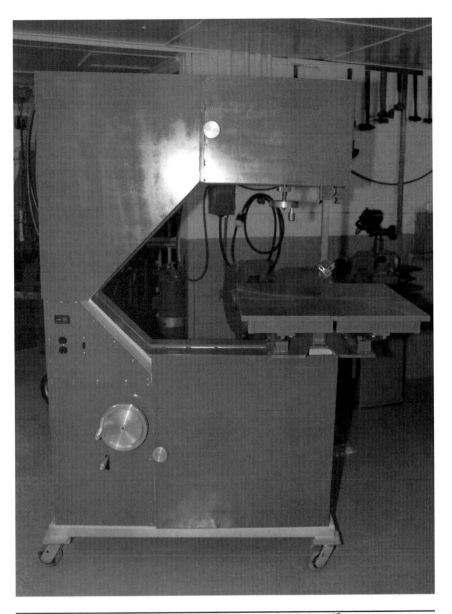

SECTION FOUR: Other Stuff

By "other stuff," I am talking about the finishing touches. The drive housing, the blade doors, a work light if you choose to install one (and you should) and, of course, painting the machine. I used my saw for several years without doors on it. It worked well and I needed to use it so installing doors became one of those jobs which required the acquisition of a "round tuit." And it took me a while to get around to it. What eventually prompted me to finish up that part of the saw was a shop cat that kept me company for a few years back in the 1990s. I had built the covered housing for the drive but I did not at that time have the plywood barrier installed between the front part of the saw and the drive compartment. I can only imagine the quickness of the cat when I started the saw one time while she was investigating the inside of the drive compartment, but she shot out of the machine at my feet with an

You know the guy who comes in the shop and says, "This will only take you a minute." You know who I am talking about. If you believe him that is your problem.

amazing demonstration of acceleration. I never found any cat hair on any of the pulleys or belts in the driving mechanism so I assumed she was agile enough to avoid them, but I installed the barrier immediately.

As a general rule I do not think it a good idea to operate any kind of a machine without all of the safety guards and shields in place. Those that make sense anyway. We are regulated these days by well-meaning rules which are necessary to protect a fool from himself and sometimes the safety features placed on machines and tools approach the ridiculous. But I suggest an exception here. This machine is so useful for making its own doors and housings that I think it may be considered a good place for the traditional exception to the rule as long as you are careful about loose clothing and you stay awake and aware of the proximity of a moving piece of saw blade with sharp teeth on it. If you have any hesitation at all about using the machine before installing the doors then by all means let your conscience be your guide.

Another justification for using the saw for cutting out the parts for the housing is this. This machine will require some "tuning" or "tweaking" before it is ready to be put to work as your everyday band saw. Belt tensioning, final pulley positions on their shafts, blade tracking and any other adjustments you may have overlooked in the construction. If you don't have to do any of that then my hat is off to you. By using the machine to complete its own construction you will have an ideal time and opportunity to make these adjustments without the guards and housings in place to interfere in the process.

I have chosen not to provide any drawings for this part of the construction. There is such a great variety of material selections and installation methods available that you will probably want to choose your own. The photographs included here should give you some ideas about how to go about making and installing the doors and covers. I am also including some thoughts on accessories you may want to build or acquire for your new saw, some of which I have built or bought and some still being thought about.

The doors may be constructed of any material you choose as long as it is sturdy enough for the job. Plywood works well but I made mine from 16 ga. sheet metal and 1/2" square tubing. The sides of the housings in the front which weld to the frame are made from 1/8" X 4" HRS strap. Cut the strap to length and weld it to the frame and weld the 1/2" square tubing to the outer edges as shown in the photographs. Grind the welds at the front so that the doors will fit flush. The doors themselves are then sawed to their final shape and a stiffening frame of 1/2" tubing is welded to their periphery. Clamp the doors into place and install the piano hinges by using either self tapping sheet metal screws or drill through and bolt with #10 machine screws.

The door knob is turned from a piece of aluminum but a valve wheel or even an actual door knob may be used. Weld a bolt to the frame in a place that will not interfere with the track of the blade and use a barrel nut to screw the knob extension into. Do

Fire trucks are red, shamrocks are green, and machine tools are gray. I hate to see machine shops that look as if they had been designed by a clothing designer. Probably just me.

1. Detail showing the installation of the 1/8" X 4" HRS strap welded onto the main frame. This also shows the piano hinge attachment to the door and the frame.

2. Upper right hand door. Notice the 3/8" — 16 bolt with locknuts used as a door latch.

3. Upper left hand door. This shows how the hinge and the 1/8" X 4" HRS strap are attached.

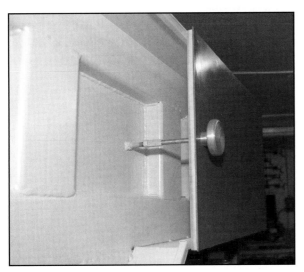

4. Upper right hand door closed and latched. The 3/8" — 16 bolt is cut so that only about 4 threads screw into the barrel nut. This makes opening and re-latching the door quick and easy.

As a weldor, the older I get the better I was.

5. Left, outside hinge installation detail.

6. Below, upper doors closed and latched.

7. Above, plywood barrier installed in my original saw. I left the pile of swarf so that you can see where it accumulates. Some sort of removable tray would be a good addition here.

8. Above right, imported blade welder mounted on the side of the original machine. This is installed close to the table so that the blade may be cut, passed through a drilled hole in the part, inside contouring sawed and then re-cut and re-welded.

9. Right, a home-made jig for silver soldering saw blades.

the same for the bottom door. Weld a strap to one of the upper doors so that one knob will serve to hold both doors closed.

The plywood barrier separating the bottom wheel compartment from the drive housing is not absolutely necessary — unless you have a curious cat! But it isn't a lot of trouble to install and it keeps a lot of debris out of the drive. It can be made from almost any material so long as it is heavy enough to be held in place with sheet metal screws or with bolts and wing nuts.

I added a Chinese-made blade welder to my saw and attached it to a swinging bracket on the side of the machine. It is conveniently located for cutting and welding the blades for inside contouring. I do recommend some sort of blade welder if you can find one. The toggle clamp shown in one of the photographs is there to assist in cutting blades to length. I have a brass screw set into the floor exactly 14' 6" from the clamp and that is how I measure blades from the rolls of blade material I use.

While we are talking about blades here is what I use. Since 95 percent of the cutting I do is steel, from 1018 to heat-treated 4140, I use a 1/4" wide, .035" thick, bi-metal blade. It has a 10/14 variable pitch. The 1/4" width is narrow enough to allow me to cut some pretty close inside radii but I guess the chief advantage of this width blade is that I can weld it on my cheap blade welder. The welder is not recommended for use with bi-metal blade material. But if I don't try to weld wider than 1/4" it seems to do OK.

For sawing aluminum plate I use a 1/2" wide, .025" thick, skip-tooth blade. The pitch, with the skip-tooth configuration, is about 1/4". This blade works well on heavy aluminum plate but you must keep some kind of lubricant handy to keep the aluminum from building up in the gullets of the blade. I use Tap-Magic® for aluminum. I apply a few drops to the blade every couple of inches of cut. This blade also works really well for sawing wood. Since it is not a bi-metal blade I can weld it on the oriental saw with no problem.

The guides on this saw will accept a great variety of blades. I won't presume to offer much advice on which to use beyond telling you what I use but there are a few important things to keep in mind. Don't try to use blades that you cannot weld yourself. Unless you come across a huge supply of ready made blades available at an auction that are of the correct length for your saw, you will not realize much in the way of economy by trying to get by without either an electric welder or some means for silver-soldering your blades. Silver solder works well, by the way, if you take the time to make a fixture like the one in the photos. The blade soldering fixture is nothing more than a piece of steel, aluminum or brass would work as well, machined with a shoulder for keeping the blade ends in alignment and a notch in the middle for the flame of the torch. Drill and tap a couple of holes for the clamps and you are in business for soldering a wide variety of saw blades. Don't tell your wood working friends about this or you will spend a lot of your time soldering saw blades for them.

File or grind the blade to a taper equal to about three times its thickness.

You are machining 3" diameter 4140 steel at about 300 FPM. You are on the other side of the saw table getting a cup of coffee when you hear the sound of a carbide cutting tool coming in contact with a hardened chuck jaw. In the 3-1/2 seconds it will take you to get back to the lathe you will hear that sound 66 more times and your shin will be bleeding. And your tool holder is broken.

10. The blade jig with one end of the blade in place.

11. Both ends of the blade clamped in the jig with a piece of ribbon solder in place.

12. The control panel. There is a start/stop push button station, a 110V duplex outlet which will be handy and a switch for the work light. Notice the channel where the blade runs.

You can bet that there have been far fewer jobs ruined by using too many clamps than by not using enough.

13. The control panel of the original saw. There is a removable cover here for the blade channel.

Clamp it into the fixture so that the tapered ends overlap each other and use a piece of ribbon solder. Clean the flux from the soldered area and if you have been careful in the preparation of the tapered ends there will not be a thick spot. If there is grind it down to equal the thickness of the rest of the blade. The half dozen or so teeth in the soldered area will not cut as well but in a 14 or 15 foot blade you will not notice it.

And don't be guilty of what many machinists will do. If you have the 1/4", 10/14 pitch blade on the saw and you need to cut a piece of 1/4" X 1-1/2" aluminum in two, it is OK to cut it with the blade on the saw. But if you have much aluminum, or wood, or TEFLON, or other synthetic material to cut, change the blade! I speak from experience here. You may think you are saving time by just using the blade which is already on the saw but it is a false economy. This machine is built so that changing blades is quick and easy so take advantage of it.

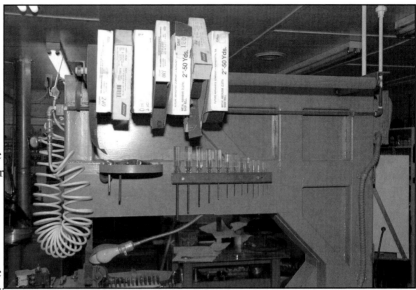

14. A photo of the back side of the original machine. A handy place for keeping rolls of abrasive paper and a rack for Allen wrenches.

One other thought about blades. Keep a couple of blades made of each type you use. I have a section of wall where I keep blades made and hanging for my machine. Murphy's law will prove out when it comes to breaking saw blades. You will break a blade when you can least afford the time to make a new one. There aren't many chances to fool Murphy but here is one.

I mentioned that on my original machine I machined a groove in the table for a miter. And now I don't use it. But you might have a need for one so don't discount the idea just because I don't use mine. Another plan I had originally for my saw was to build a trunnion for the table mount so that it could be tilted from side to side for sawing angles. This would be an interesting project but I just never got around to doing it. There have been a couple of occasions over the years where I would use this feature if I had it but I didn't so I went in another direction.

Vendors who will tell you that they are selling a system that is "foolproof" have never met some of the people I have met.

There are other convenient accessories you may want to consider. A piston air pump operated by an eccentric installed on one of the drive shafts for blowing chips away from the cut. I have used saws with this feature and it is not as frivolous as it sounds. I sometimes use an air hose to do the same thing. Some type of an easy to empty chip catching hopper would be a good thing. A speed indicator can be made by using an electronic tachometer to read the RPM of one of the pulleys or of the bottom wheel and then making a chart to convert RPM to SFPM. A gravity assisted feeding device will be useful if you plan to do much sawing of heavy materials. As I sit here writing this I am realizing that I have still not finished with my first bandsaw! But there will come a time when you will

15. Right, a toggle clamp attached to the base of the saw. There is a brass screw installed in the floor exactly 14' 6" from the edge of the mounting block. For cutting new blades from a roll of saw blade material.

16. Below, an example of what this machine will do. These are parts sawed from 1" thick 4140 alloy steel plate.

find that you have put enough time into the building of the machine and now it is time to use it.

Paint the machine whatever color you prefer but my philosophy is that fire trucks should be red, toilets and urinals should be white and machinery should be gray. But paint it. It will perform just as well unpainted as it will painted but after all the time you have invested in this project you will surely want to finish it off by painting it! I know a machinery rebuilder who would probably get out the Bondo® and the sandpaper and make a show piece of a machine like this. There are practical limits however and you will have to determine where the limits lie.

A bad day spent in my shop with my tools is better than a good day spent almost anywhere else.

As I said at the beginning of this project it is not a weekender! It might be with enough weekends but there is a lot of time involved. If you have taken the time to build this machine, or some modification thereof, you will not be sorry you did. I use my milling machines once or twice a week. I use my lathes probably an average of four or five times a week. I only use my surface grinders and my heat treating ovens maybe once every two weeks or so but I use my bandsaw almost every day. I am proud of it. I show it off to people who visit me. And now I have two of them! Anybody want to buy a good vertical bandsaw?

Chapter 3

Removing Broken Taps and Studs by Welding

There are a couple of ways of looking at this. Some think that the person who enjoys doing for a hobby the same thing he does for a living is some kind of a nut. Personally, I think that person is just about as lucky as anyone can possibly be.

This article was printed in its original form in the March/April, 2004 issue of *Home Shop Machinist* magazine. It was the first of my work accepted by that excellent magazine and the fact that it was accepted probably gave me added incentive to do the book you now are holding. Since its original presentation I have added the section about removing broken dowel pins as well as adding some additional photographs of incidents where the process described has worked for me. There have been times when the process did not work. You will notice that I do not include any photographs of those adventures nor do I spend a lot of time talking about them.

In the early 1970's it was my good fortune to work in a machine shop with Smitty. Smitty at that time was approaching 80 years of age and has since passed on but he took with him more common sense and machine shop experience than it seems fair for one man to have. He gained his early experience in the railroad shops of the era before the depression and I don't think he ever forgot anything. When I met him he only worked half a day at a time and was mostly in the shop to provide the rest of us with the benefit of his experience. Smitty could make a job easier by just hanging around while the work was being done.

1. Samples of parts removed with this method.

I was building a steady rest for a 5L Gisholt® turret lathe from 6" A36 plate and was tapping a series of 1/2"-13 holes when I broke a bottoming tap off in the hole. Smitty looked up at me with a grin and said, "I never broke a tap in my life..."

I was thinking of the best and most respectful way of calling him a liar when he continued, "...that wasn't a %##&*& to get out!"

He made his point. But this tap proved not to be so much of a %##&*& to remove because Smitty showed me a trick and here it is. This is a method that works for some broken bolts or taps. It doesn't work every time but I have found that it is usually worth a try and when it does work it is satisfying. If it doesn't then all I have lost is a little time.

Photo 1 shows some examples of parts I have removed with this method. From left to right you will see; (1) A 1/4"-20 flat head cap screw which was frozen in the part and the corners in the hex socket were rounded off. I welded a 5/16"-18 nut to the head of the screw and removed it. (2) A 5/16"-18 screw which I removed in the same way. Notice that this screw has a left-hand thread which was why it was twisted off. (3) A 3/8"-24 set screw with a 3/8"-16 nut welded to it and (4) A 1/2"-20 stud removed, also by welding to a 3/8"-16 nut.

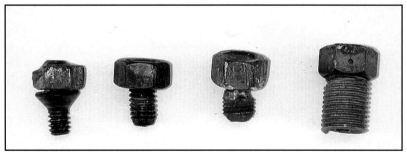

2. Broken tap on left, with three tries to remove it.

3. A stud intentionally broken to demonstrate process.

Photo 2 shows a series of three tries that it took to remove a 3/8"-16 helical tap from a mild steel lapping ring for a customer. The part was brought to me to drill and tap (6) holes around a 24" bolt circle. I was, naturally, on the last hole when the tap broke. On the left is the remains of the tap. Shown next is the first attempt. This resulted in bringing out just a fragment of the tap which was probably broken when the tap shattered. The next attempt did the same thing but brought out a little more. By this time I was deep enough into the part to run the risk of welding the tap to the part so I made a sleeve by drilling a 1/4" hole through a 3/8" bolt and screwing it into the part until it reached the top of the broken tap. As you can see it went down to a depth of about 1/8." I then reached down through the nut and the sleeve and welded both parts to the tap. There was much celebrating and self-congratulation when the remainder of the tap came out of the hole. I used 3/8"-16 nuts for all three attempts.

Photo 3 shows a stud I have intentionally broken off to try to demonstrate this process. The stud is a 1/4"-20 steel bolt 1/2" long and is broken off flush with the surface of the hub of the taper-lock pulley. It is good practice to always use the biggest nut you can without running the risk of it being so large that you weld the nut to the part rather than to the stud or tap you are trying to remove. In this example I used a 1/4"-20 nut. You can also make your own nut or part to weld on by drilling an appropriately

4. In position to weld the nut on.

sized hole through a piece of square stock or even round stock. If you don't have flats to engage a wrench you can use vise grips to turn the stud out.

In Photo 4, I am in position to weld the nut on. I used a 1/16" stainless steel electrode which I keep a supply of for just this purpose. The electrode is 308L-16 and works best at about 65 Amps DCEP or reverse polarity. The L in the electrode number indicates low carbon which is not necessary for this purpose. Almost any electrode will work here. Just keep in mind that it must be small enough to establish an arc in the hole and then weld to the nut without too great a metal deposit. Stainless steel works well because it is easier to strike an arc with just a touch and it welds at a lower heat. Photo 5 shows the weld being made.

After welding, tapping the nut a few times with a hammer can help to loosen studs that may have been rusted in. The heat of welding also contributes to the loosening of the part. In Photo 6 you see the nut welded on and Photo 7 shows the stud successfully removed.

It is sometimes necessary to make a sleeve, shown here in Photo 8, to help to reach down for a tap or stud broken off below the surface. This

5. (Top) The weld being made.

6. (Right) Nut welded on.

works best if you make the sleeve to the same thread as the broken part. Try to insure that the sleeve reaches all the way down to the top of the part you are attempting to remove. The series of Photos 9, 10 and 11 show this part being welded and removed. You can see the juncture of the sleeve and the imbedded stud at the arrow in Photo 11.

7. Stud successfully removed.

8. Sometimes it is necessary to make a sleeve.

9.

9, 10 and 11: Part being welded and removed.

10.

11.

Ignorance is one of the most underrated characteristics a work- man can have. If you don't know that some- thing cannot be done you might just do it.

12, 13, and 14: Part being welded and removed using MIG.

I am most accustomed to the method I have described here but Photos 12 through 14 show the same result accomplished with MIG. It depends on which welding system you are most comfortable with and, of course, on what you have available. For this example I used .030 wire and a 75%/25% mix of CO_2 and Argon. Wire speed and voltage settings will vary with your machine.

There will be times when you will be asked to remove non-threaded items such as hardened dowel pins which have been driven into blind holes. This method will work for those as well but you have to approach it a little differently.

12.

13.

14.

Photo 15 shows a work-holding chuck jaw from an Okuma Howa turning center which is owned by one of my regular customers. There were four of these jaws with broken 1/4" dowels in them, two each in three of them and only one in the fourth. I was successful in removing all of the broken parts by using the welding process. I started in a couple of cases by building up the end of the broken dowel with the same 1/16" stainless steel electrode I used in Photo 4. I built up the end of the broken dowel until I had enough material in place to weld a nut to. In Photo 16 you see an example of where the weld ran off onto the part

15. Chuck jaw with broken dowel.

16. Showing where weld ran off onto part being worked on.

17. Using a file to remove weld run-off.

Just when I get used to referring to a process by its given name somebody will change the name. What was once Heliarc became TIG. I was beginning to like that when it was changed to GTAW. I can almost tell you a weldor's age by what he calls what he is doing.

being worked on. (I am not as steady as I once was but you should have seen me in my youth!) Use a file to remove the area as shown in Photo 17.

After welding the nut on, I then welded the head of a bolt to the nut. I then stacked enough spacers, 1/2"-13 hex nuts worked well here, to provide clearance, placed a washer and a nut onto the welded-on bolt and pulled the dowel from the hole. Photos 18 through 23 show the steps required to accomplish the job. Use the biggest parts you can conveniently use here. Nuts to be welded on, draw bolts to be used, etc. You will be limited by the size of the parts being worked on but don't hamstring yourself unnecessarily.

Photo 24 is of several other instances where the described process has been useful. Notice the drill bit in the center of the photo. It was really a pleasant surprise to me when it came out as easily as it did because of the depth it had reached when it broke off. I have a small box full of these examples now and I have no idea why I have saved them unless it is to drag them out from time to time and brag about them.

One additional comment I will make here is about the welding processes involved. I have used in all of these examples either the SMAW or GMAW process. There are times when GTAW may provide better results, particularly when the parts are small or especially sensitive to heat. There are no universal fixes, either in removing broken parts from deep holes or in any other area of our lives. But we keep on trying.

As I said at the beginning of this article, this doesn't always work. If it doesn't work the first time it is usually worth a second or

18-23: Series of photos shows steps required to finish job.

third try. I have used up as many as half-a-dozen nuts before getting the results I want but nuts are cheap compared to the expense of other methods of removing broken taps or studs.

20.

21.

The farmer, the mechanic, the plumber, the carpenter, even the golfer, all have one thing in common. They all know where the guy lives who has tools and fixes things.

22.

23. Final step in process.

24. Examples of other instances in which this process has been useful.

Chapter 4

Building a Rifle Rest

Building this rifle rest was inspired by my seeing one my uncle had built in his shop many years ago. I show it here in Photo 1. My uncle also has built a number of rifles, including the one shown here in the cover photo for this chapter, and he has always been one of the people whose talents I most admire in the machine shop. He could build or make anything he wanted. Anything! I have described his abilities in this way. He is at least twice as good as the next best machinist I ever knew. He is a master of the material and the machine. I would like to think I could someday be as good.

This construction article appeared in the June/July, 2004 issue of *Machinist's Workshop* and was nominated by a member of *The Home Shop Machinist & Machinist's Workshop* BBS forum as the best construction article of 2004. The nomination was seconded but it never came to a vote so as far as I know those two people were the only ones who liked it. Thanks, J and Cecil!

Marksmanship, like wealth or beauty, means different things to different people. It has one definition for the skeet shooter but means something entirely different for the cowboy action shooter. For the bench rest shooter it means putting all the bullets through the same hole in the target. Or as near to that as is possible. I make no guarantee as to whether the rifle rest described in these lines, pictures and drawings will improve a shooter's marksmanship or just make it more difficult to carry all the accessories from the trunk of the car to the shooting bench but it won't impair your shooting and it is fun to build. And it gets a lot of notice at the local shooting range.

I approached the selection of materials for this job as I do every job. I applied all of my engineering experience and my knowledge of metallurgy and machining processes. I gave careful consideration to weights and to the effects of welding and machining on the various possible choices of material. And then I went to my scrap bin, or, as I have noted elsewhere in this book, to my secondary materials storage facility, to find out just exactly what I was going to use for the job. I did choose heavier sections for this job. A heavy rest provides stability and inertia to the user and those factors will contribute more to the consistency of your shooting than almost any other variable. Excluding the wind, of course.

The materials list shown here is provided only as a guide. You will know where and when to make substitutions from your own secondary materials storage facility.

We begin by sawing the three legs, Part number 08, to an

Sometimes lighting an oxy-acetylene torch with a butane cigarette lighter can be more entertaining than you would like for it to be.

1. Rifle rest built by my uncle, George B. Bulgin.

MATERIALS LIST
Base and legs

1. 1" X 1-1/4" X 6-1/2" Rectangular HRS bar. Part number 08. (3) required for legs.

2. 3" diameter X 3" long CRS steel shafting. Part number 01. Center housing.

3. 1-1/2" diameter X 16" long CRS. (Includes material for feet and locking nuts). Part numbers 06, 09 and 10.

   Parts Nos. 01, 08, and 10 should be mild steel for welding. Either HRS or CRS will serve.

4. 5/16" X 1/2" X 3" long key steel.

5. (2) 1/4"-20 socket head cap screws 1/2" long.

6. (3) carbide chuck grippers. Reid Tool Supply Co. catalog number RG-825. *

7. Bulls eye level. McMaster Carr Supply Co. catalog number 22325A11 *

   *Optional

Spindle, saddle and elevating nut

1. 4" diameter X 1" long bronze or aluminum. Preferably bronze. Part number 05.

2. 2-1/2" diameter X 5-1/2" long CRS. Part number 02.

3. 1/2" diameter X 21-1/2" HRS steel. Handwheel for part number 05.

4. (3) 3/8"-16 bolts 3" long. Spokes for handwheel.

5. 4" X 8" X 1/4" wall thickness rectangular tubing 6" long. Part number 07.

6. (1) 3/8"-16 X 1" long Socket Head Flat Head Screw. Attachment screw for Part number 07.

7. (1) 3/16" X 3/4" long dowel pin.

8. (2) pieces 1" X 1" square HRS 1" long and 1-1/2" long. Part numbers 03 and 04. *

9. 3/8" diameter CRS or drill rod. Forearm stop parts. *

10. (2) Knurled thumb screws. Reid Tool Supply Co. catalog number AJ-539 *

    * Optional parts for forearm stop.

approximate length of 6-1/2." I then used the setup shown in Photo 2 to cut the 13 degree angle on the ends of the legs and to insure they are cut to the same length. The taper of the legs is necessary only for appearance as is the 5/32" radius machined on all four corners of the parts. I machined the radius with a corner rounding end mill but the same effect may be achieved by machining a bevel or even by grinding a radius or a bevel after welding. A square corner may serve your purposes equally well. The weld preparation is ground on both ends of the legs. A 100% weld is not necessary here but

2. Sawing the angles at the ends of the legs.

it must have enough depth of weld to insure there is no danger of grinding the weld away while dressing up the job.

Photos 3 and 4 show the taper and the radius being machined on the legs.

The parts required for the base weldment are shown in Photo 5. It is an almost universally accepted rule in steel fabrication that all of the welding should be completed before beginning the machining processes required for the job. I agree with this and I try to apply it to any job I do which requires both welding and machining. But sometimes we have to make compromises. We cannot all have access to a VTL or to a bar mill and many times the completed weldment is simply too large for the machines available. This is where we sometimes have to get innovative. If you cannot swing the completed weldment in your lathe as I did in my 17" lathe then here is a suggested alternative route to follow.

3. Below, set up in the milling machine to machine the taper on the legs.
4. Right, machining the radius on the corners of the legs.

Machine the center housing, Part number 01, to its finished OD. Bore the ID but leave about .015" to .030" material in the bore. Machine the annular lubrication groove as shown in Photo 6 in the top of the part and remove it from the lathe. After welding on the legs, the bore

is likely to be distorted slightly from the heat induced in the welding process and we will correct this by setting the part up in the milling machine and taking a light cut through the bore. The radial lubrication grooves may be laid out with a straight edge and machined in the milling machine at the same time. A rotary table will give you more accurate location of the radial grooves but accuracy is not a concern here.

The welding jig shown in Photo 7 is important. There are many ways of doing this and I show here only one of them. I laid out the pattern on my welding table and welded bolts and angle iron clips directly to the table. You can use a piece of 3/4" plywood but you will, of course, have to connect your grounding cable directly to the job. Photo 8 shows the parts clamped into place and ready to weld. Tack all the joints in at least two places each so as to minimize distortion. After tacking, remove the base from the jig and complete the welding. Photo 9 shows the parts as welded and Photo 10 shows the base after blending the welds with a hand grinder. A word here about finish grinding. If you have not yet discovered flap discs then stop what you are doing right now and go find a supplier where you can buy flap discs to fit your 4" hand grinder. I use 40 grit which is useful for the aggressive removal of material and is unsurpassed in blending welded contours. The 1/4" bolt with large area washers through the threaded holes seen in Photo 9 in the legs are there to protect the threads from weld spatter. It is a good idea here to run a tap through the threaded holes after welding to insure there is no distortion. These threads need to be a good fit because you don't want to have to get out the pipe wrench every time you need to change the elevation of the rest.

The 2" keyway through the bore of the center housing does not have to be a full depth keyway. It serves merely

5. Left, all of the parts to be welded.
6. Above, machining the annular lubrication grooves.

7. Below, welding jig.
8. Bottom, clamped up for welding.

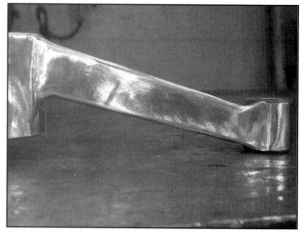

9. Above, welded before grinding.
10. Right, welds ground and blended.

as a guide for installing the key which is bolted into place with (2) 1/4"-20 socket head cap screws. If you do not have a convenient way of cutting internal keyways of this size a 1/2" bolt threaded through the side wall of the base and protruding 1/4" into the bore will serve as a key. Its purpose is merely to prevent the internal screw from rotating while in use or while adjusting the height of the rest.

The leveling screws and locking nuts, Part numbers 06 and 09, are machined from 1-1/2" diameter CRS. Turn the threaded legs to .625" in diameter and thread for 5/8"-18. A full length thread is necessary here so a relief groove must be machined just below the head of the part. Machine the parts to completion before parting off from the bar by drilling and tapping the 1/4"-20 thread in the end and knurling the head. Drill, thread and knurl the locking nuts from the same material. I much prefer the knurling tool shown in Photo 11 because it produces a more consistent knurl and greatly minimizes the lateral thrust on the job. It is easier on cross slide screws and, in smaller lathes, does not apply such a great pressure to the headstock bearings.

I installed

11. Knurling the leveling screw.

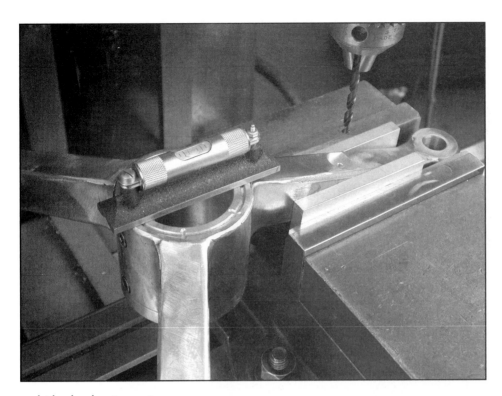

12. Leveling the part for machining the socket for the Bulls-Eye level.

carbide chuck grippers in the bottoms of the feet for this rest by drilling and tapping the 1/4"-20 holes. They are not really necessary for the efficient function of the rest but if you shoot from concrete shooting benches they may be useful. They are available from Reid Tool Supply Co. as Cat. number RG-825 for $5.85 each plus S&H. An alternative is to silver solder in a piece of a carbide drill bit, or even a short piece of HSS, and grind to a point.

I will not pretend to you that the bull's eye level in the leg of the rest nearest to the shooter is necessary or even

13. Completed base before painting.

useful. But I thought it added a nice touch and it isn't all that expensive. The level you see here is from McMaster-Carr and is their part number 22325A11. It cost $8.20 plus S&H when I bought it in 2001. If you choose to install one, set up the base as shown in Photo 12. Center the spindle of the milling machine at the approximate midpoint of the length of the leg and centered to its width. Machine a flat with a 3/4" end mill and then

14. Above, drilling the elevating nut with a spade drill.

15. Below, machining the Acme #8 threads.

drill through with a 1/4" drill. Counter drill to a depth of approximately 1/2" with a 47/64" drill and then counter bore with either a flat bottomed 7/8" drill or a 7/8" end mill. Push the level into place after painting the base. This setup is also the time for adding the radial lube grooves in the top of the housing if you haven't already done so. Photo 13 shows the completed base before painting.

We begin the elevating screw and saddle for the rest by machining the screw, Part number 02, to its finished diameter of 2.125." Machine a relief groove at the top end and machine No. 8 Acme threads the full length. Finish this part by machining a 1/2" X 1/4" keyway the entire threaded length of the part and then drilling and tapping the 3/8"-16 hole in the top. Drill

a 3/16" hole 3/4" from the center of the part for a locating dowel pin.

The elevating nut, Part number 05, was machined from a piece of 4" diameter AMPCO 18 aluminum bronze alloy 1" in length. AMPCO 18 is a tough, long lasting alloy used for the manufacture of worm gears and other parts where severe wear is a factor. I used it simply because I had a piece of it. Any brass or bronze material will serve equally well here and even a piece of 6061 tempered aluminum will work. The nut should be made of a material dissimilar to the spindle so as to minimize friction and wear. AMPCO 18 will probably last in this application for a minimum of a thousand years or so. Photos 14 and 15 show the part being drilled with a spade drill and machined for the No. 8 Acme thread. Notice that

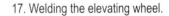

16. Setup for drilling and tapping for the elevating wheel spokes.

17. Welding the elevating wheel.

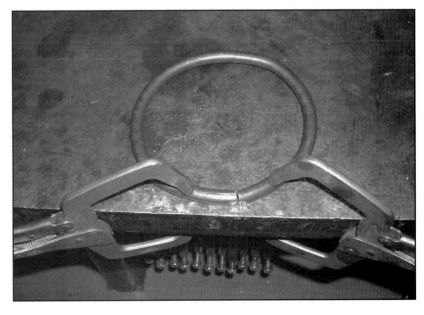

for drilling the part is placed so that the body of the chuck takes the thrust of the drill. It is then moved out to the front of the chuck for finish boring, facing and threading.

Photo 16 shows the nut set up on the milling machine in the rotary table for drilling the holes for the spokes. Drill and tap these holes for 3/8"-16 to a depth of approximately 1/2." Cut the threads on (3) 3/8" bolts to a length of less than 1/2" and screw them tightly into the holes in the periphery of the bronze nut. They will be cut to length when you are ready to weld them.

Form a ring from the length of 1/2" CRS. This can be done cold but if you have access to a small forge it will make it easier. Bring the two ends together and clamp to the welding table as shown in Photo 17. Weld the joint and then true up the ring before welding in the spokes. Photo 18 shows the ring, spokes and nut in place and ready to weld.

The rifle tray, Part number 07, is a straight-forward weldment made from (3) pieces of 1/4" plate or, preferably, sawed from a scrap piece of rectangular tubing with a 1/4" wall thickness. The tray is mounted to the spindle using a single 3/8"-16 socket head countersunk screw with the 3/16"

18. Top, ready to weld in the spokes.
19. Above, locating dowel.
20. Right, forearm stop parts.

I have never actually dropped a refrigerator into the chip pan of a lathe. But I will bet that if I ever do I won't be able to find it without a flashlight and a chip hook.

diameter dowel pin serving to lock it against rotation. Photo 19 shows the relationship of the center mounting screw to the dowel. Be sure that the dowel is in the correct position relative to the axis of the saddle.

The forearm stop, Part numbers 03 and 04, is a simple addition shown in Photo 20. It serves simply as an adjustable stop to position the forearm of the weapon being fired in the same place for each shot. A plastic cap from one of the tubes end mills are packaged in will serve as a guard against marring the finish on the forearm of your pet rifle.

Fill the lube grooves with a bearing grease thick enough to not run out and get all over your trigger finger and then go the local range and be prepared to answer questions about where you got the rest. You will be asked. The finished rest is shown in Photo 21 minus the bags and the rifle.

I have a friend who does upholstery work and I hired him to fabricate the sandbags from leather and I filled them with dried beans. There are several options here and those of you who do much shooting will know of sources for factory made sandbags. In a pinch any cloth sack filled with sand from a construction site will serve but after spending the time you have spent building this rest you will likely not be satisfied with that. As a matter of fact, you can buy a rifle rest and save yourself several hours of fabricating time. But then, that isn't the way machine shop hobbyists do it, is it?

I admire the process engineer who will go to the shop floor and talk to the operators when writing specifications for a new machine tool. And I don't have much use for the one who will not.

21. Completed rest.

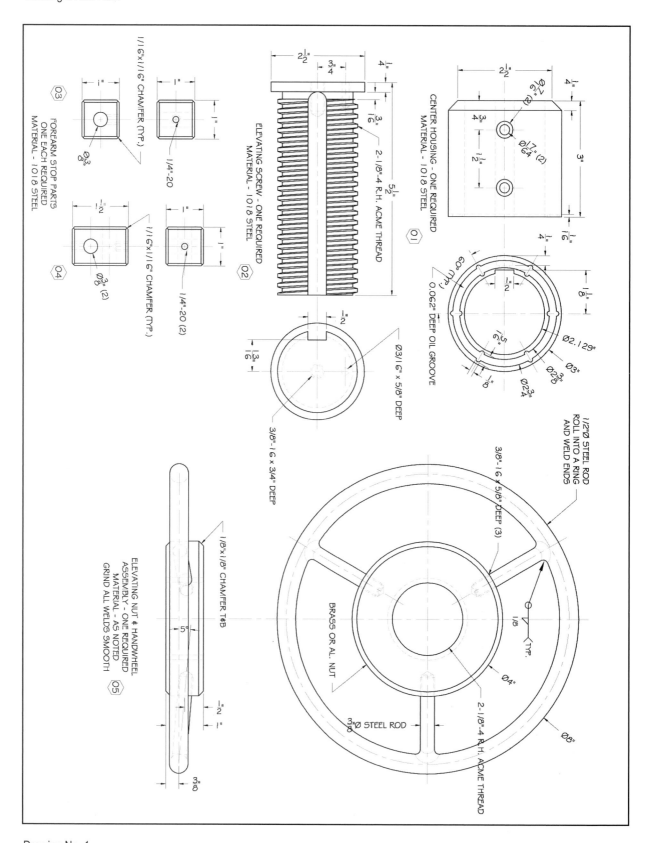

Drawing No. 1.

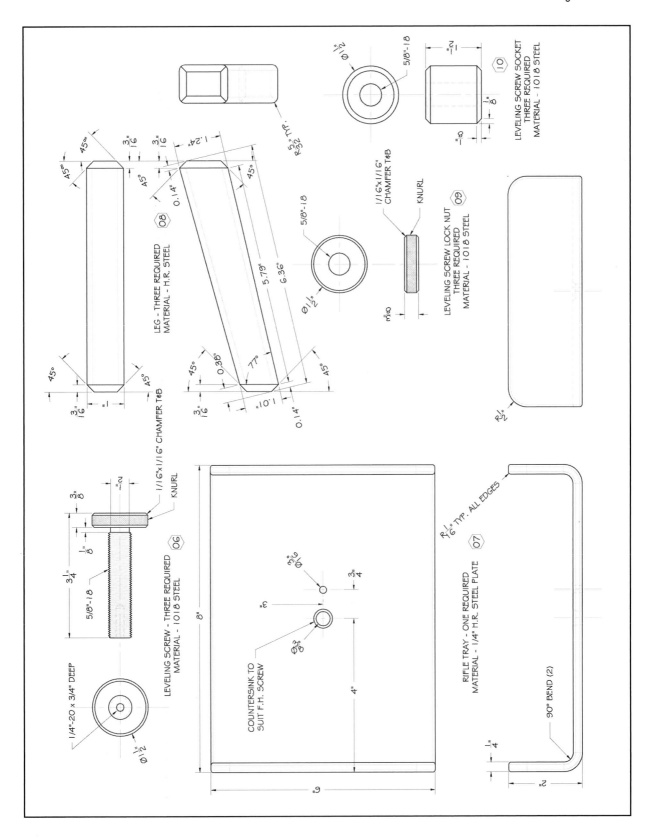

Drawing No. 2.

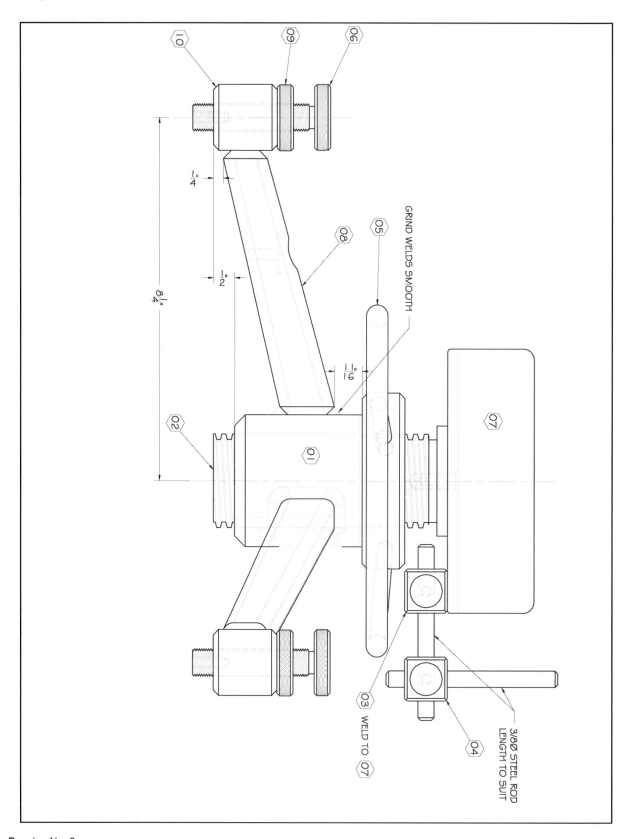

Drawing No. 3.

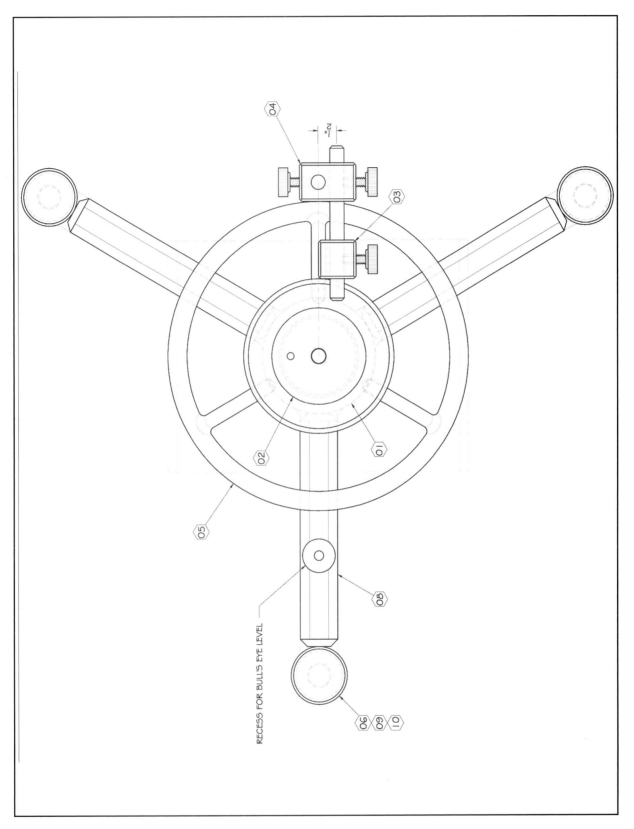

RECESS FOR BULL'S EYE LEVEL

Drawing No. 4.

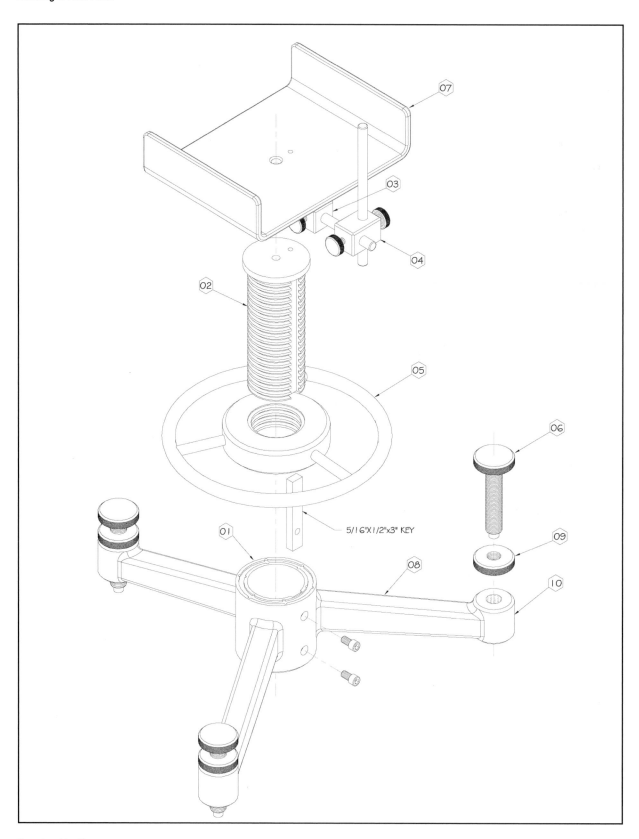

Drawing No. 5.

Chapter 5
Lifting Devices for the Small Shop

How an insert or a small part can fall straight down into the chip pan and still end up 30" away from where it fell is beyond me.

A. Changing chucks on the Kingston lathe.

I have hesitated to include this chapter in this work and here is why: cranes and other lifting devices are responsible, directly or indirectly, for more injuries in workplaces where these devices are used than all of the other causes put together. Excluding minor cuts and burns from chips and sparks, of course. But the fact that they present a danger does not make them inherently dangerous. As is the case with every tool in your shop or mine, any danger of injury associated with the use of these devices depends almost entirely on you or me and on how we use the tool. Or the fork lift. Or the crane. Or the jack. But sometimes we have to have them so let's talk about them.

And now here is my caveat. The lifting tools and devices shown in the photographs accompanying this chapter are all in use in my shop. They were all designed and built by me and they all do the job they were designed to do. But I do not recommend any of them for anyone else. If you have the talent and the ability, and many of you reading this do have, to design and build your own cranes and hoists then go for it. But if there is the slightest doubt as to whether a component will handle the load you are considering lifting then you must consult with an engineer and listen to what he says. And by components I mean every piece of the construction from the floor to the sling! A bridge crane with the capacity to hoist one thousand pounds, which is really quite a small crane, built to exacting specifications as to columns, rails, bridge, hoist trolley, hoist, cable and hook is of no value to anyone if it is mounted to a 2 X 4 framed wall with 3/8" gypsum board for wall covering. Of course you would not build something like that. But I stopped being surprised years ago at the foolishness some people are capable of. So... having said that I will say this:

• Do not build any lifting device unless it has been properly

engineered as to choice of materials, fabrication and construction, and of course, installation.

• Do not use slings, chains, ropes, cables, hooks nor string if there is any doubt about its ability to lift the load being handled.

• If it is necessary to weld on assemblies suspended from a hoist make ABSOLUTELY sure that the grounding clamp is properly connected so that welding current does not pass through lifting components or bearings.

• Never work under suspended loads without first installing adequate safety supports or jack stands.

B. Moving the indexing head from its storage place. Notice the storage rack for the lathe chucks on the wall. The centerline of the storage rack is the same height above the floor as the centerline of the lathe spindle.

Now then. If it is perfectly clear that these comments and photographs represent suggested solutions only for your shop's lifting needs take a look at some of them.

When I think of a bridge crane the first thought that comes to mind is the crane which served the shop I worked in as a supervisor for ALCOA. It was a 25 ton crane, a Cleveland I think it was, and was installed in the North Plant machine shop when the factory was built in 1940. It had an open cab which was hot in the summer and cold in the winter. You gained access to it by climbing a vertical ladder and typically, the operator came down one time during an 8 hour shift for a lunch break or a visit to the restroom. The

C. (Also shown on chapter cover) Handling a large diameter job in the lathe. The assembly, workpiece, chuck and adapter plate, weighs a little over 200 pounds. Tough to handle without the hoist.

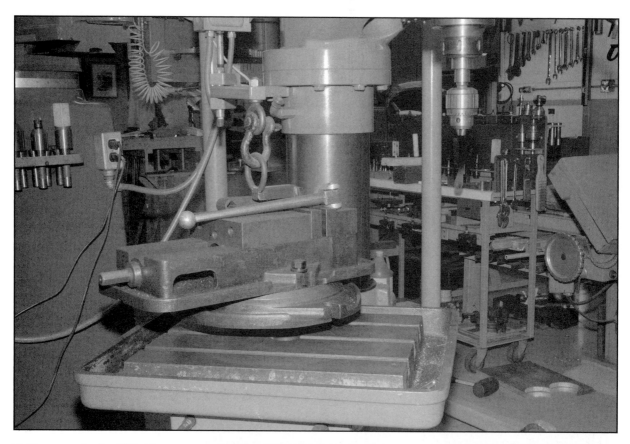

D. Above, moving the 8" Kurt in place on the drill press.

E. Right, the 8" Kurt in its home position.

bay served by that crane was the center bay in the shop and, if memory serves me, it was 75 feet wide and about 500 feet long. When a machinist or a mechanic needed the crane the procedure was to pull a lanyard which blew an air whistle and the crane came to you. The two ladies who operated the crane on the two shifts from 6 a.m. until 10 p.m. were so skilled that the hook, moving simultaneously in all three dimensions, arrived at the point in space exactly where it was needed without swinging an inch. It just

F. Outside gantry in position over the sandblasting platform.

G. Inside gantry over the cutoff saw. Also serves the welding table.

came to a stop at the palm of your hand ready for you to attach your slings or cables or whatever else you might want to place on the hook. It was a joy to watch anyone so experienced and adept at their job. Most of those old time crane operators have been replaced in recent years by remote controls.

I have worked under many other bridge cranes and polar cranes in my lifetime of shop and construction work. A polar crane for some of you who may never have seen one, is a bridge crane which operates in a circular building. The bridge rotates instead of moving in a linear direction. Bridge cranes and polar cranes located in power houses and in nuclear reactor buildings can have the capability of lifting up to 500 tons and more. But having a crane with that capacity does not mean that every lift you make will be a 500 ton lift! The bridge crane I have over my lathes and

One of the most conscientious people I ever knew about wearing safety glasses was a one-eyed turret lathe operator.

H. Swinging jib crane with "jill poke" or outer support in place.

It is amazing what about three quarters of an inch of rubber mat can do when it is between your feet and a concrete floor.

milling machine has a capacity of 1,000 pounds. I tested it at that when it was built with a 1,020 pound welding table. Lifting that sort of load, however, is not what it was built for. It was built not to lift the big load but to lift the small load many, many times. The four-jaw chuck for my 17" Kingston lathe weighs 90 pounds. The three-jaw chuck weighs 67 pounds and although these are not huge accessories they are bigger than I want to wrestle around while trying to reach the chuck wrench which will typically be lying about 4" beyond my reach after I get the chuck in place on the spindle. I drilled and tapped into both chucks a hole for an eye-bolt and now it is a simple matter to hook onto the chuck at the storage rack, which is built at exactly the same height as the lathe spindle, by the way, move it around to the front of the spindle and put it into place.

The same hoist, which is a simple one-half ton chain hoist, serves to place the 74 pound 6" Kurt milling machine vise or one of the 125-pound indexing heads on to the milling machine table. Those accessories are stored on my one-wheeled carts which will be explained in more detail when we talk about space management in Chapter Nine. Handling these tools is easy and convenient and hardly a day goes by that I do not change chucks or milling machine accessories.

Installing heavy accessories on machine tools is one use for this little crane, and in my shop, the primary use. Frequently I will have a job to do on one of the machines which is made easier by the presence of some sort of lifting mechanism. It doesn't take but 30" of 4" diameter CRS to get up to about 100 pounds and it is far easier to put a piece of stock like that in the lathe when you have an overhead lifting device. In one of the photographs included here is a job I did which called for removing the gap from my

I. Storage place for "jill poke" when it is not in use under the jib crane.

gap bed lathe. I bolted the circular piece of plate to an adaptor plate which could in turn be grasped by the four-jaw chuck. The chuck, plate and the adaptor plate probably weighed in the neighborhood of 200 pounds and I had to machine both sides of the plate so there were a couple of times that it had to be handled. It would have been a really difficult task for a broken down old machinist like me to handle a job like that without having some sort of lifting assistance. So there are times when this little crane has been responsible not only for making my job easier, but for allowing me to take the job at all

Not all lifting devices are cranes. My drill press is not under my bridge crane (the drill press pokes through the ceiling of my shop by about 4" making it difficult to build a traveling crane which will clear it!) and I have an 8" Kurt swivel vise which has to be placed on the table and then removed from time to time. The vise weighs over 200 pounds so I had to come up with a different system. To remove the vise from the drill press table I swing the table around as shown in Photos D and E, elevate the table until the vise can be clamped on to a bracket and then lower the table away from the vise. The bracket supporting the vise is then swung away to its storage position at the side of the drill press and the table brought back to the front of the machine. The entire system didn't take more than an hour to build and install and it has probably saved me from a big bill from the orthopedic clinic! Not to mention the mashed

J. Jib crane in use.

K. Columns and cross bracing in place for building bridge crane.

L. Crane rails and bridge in place.

fingers which can result from handling this type of tooling accessory.

I have two gantry cranes, one outside over the sandblasting platform and table and one inside which serves the welding table and the cutoff saw. If I have items large enough to warrant it, I will move machines and pieces of machines from the shop to the sandblasting table and back with the help of the trailer. The inside gantry moves materials from the hand truck to the saw table and back to the hand truck. I buy CRS in diameters up to 6" and I usually will purchase a piece three feet in length. That weighs in at close to 300 pounds so I need to be able to have it under some sort of hoist all the way from the delivery truck to the cutoff saw. The various systems I have built enable me to handle material of this size. The point of all of this discussion about gantries and hand trucks and wheeled trailers and dollies is this: everything in my shop which has to be moved on a regular basis, which includes about everything in my shop except the larger pieces of machinery, is either

M. Left, welding support bracket for crane rail.
N. Above, mast for jib crane shortened to allow the passage of the bridge.

on wheels or can be lifted and carried to wheels. I am too old and far too lazy to be lifting materials and tools which weigh more than my dinner bucket.

The most recent hoist at my shop is the bridge crane which is shown both while under construction and after completion. At one time a jib crane was the only way I had to unload machines at the back door and it has

O. Lots of lifting capability. Swinging jib crane in foreground and traveling bridge crane in background.

unloaded a lot of treasures in its day. But it was limited. At the extreme end of the boom I would have been afraid to try to lift more than a couple of hundred pounds without some outer support so I built the "jill-poke," to use one of our country expressions, to support the outer end. I unloaded my Kingston lathe with that system which was probably all the little hoist wanted to handle. I knew that I needed a little more lifting capability back there so I built the bridge crane which then, of course, called for building a

P. Detail showing installation of crane rail support bracket.

roof over it which then, of course, called for constructing some sort of walls around it which will now have to be wired for three phase and plumbed for air and so it goes. There is no end to it and I don't want there to be. After finishing the bridge and before adding the roof, I cut the jib mast down so that it can swivel under the bridge and it now serves to load barrels of chips and the small lifts into and out of my pickup truck.

The final example I show here is of the rail and trolley in the back room. The rail is fixed in position and the hoist does not have a lot of weight lifting capacity but it is frequently handy for moving jobs in and out of the back door. It will probably be the subject for the next upgrade of my lifting systems.

This list of material handling accessories would not be complete without including a description of my trailer. I have the bad habit of frequenting machinery auctions and have been fairly successful as a one-man show in buying, reconditioning and then re-selling small to mid-size machine tools. I imposed on a friend for a long time by borrowing his trailer to move my treasures from the place where they were purchased to my shop. I was tired of having to borrow a trailer every time I had to move a machine and I know he was tired of loaning it so I built the trailer you

Rule of thumb — keep it out from under the hammer.

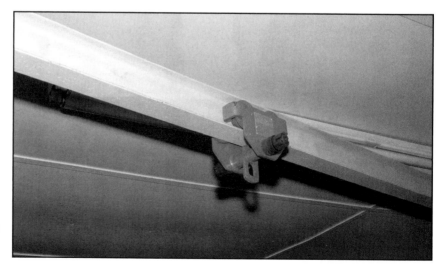

Q. Short section of 20' long fixed beam hoist in back room of shop.

R. Flat bed trailer. Notice hand cranked winch mounted in front.

see here in the photographs. I built a set of roller ramps for the trailer and now, with the aid of the small, hand-cranked winch mounted at the front of the trailer, I can load some pretty heavy machines without assistance. I can now bring a machine, or a part of one, to the sandblasting platform where it can be unloaded with the gantry located there. After sandblasting, or pressure washing as the case may be, I reload it onto the trailer and move it up to the new hoist in the enclosure at the back of the shop where it receives whatever other attention I may want to give it. And by the way, I would have had trouble building this trailer without the presence of the jib boom at the back door.

These examples are just a few of the ways you can make life easier in your shop. They were not all built at the same time but rather as they were needed over a period of several years. The first thing to be done is to identify the need. You will then think about it for a time while you start accumulating the necessary materials and during that process you will revise in your mind what the final product will be. Then after you build it you will probably see things you would have changed and you will call those things back to mind when you make the next addition. It is an evolutionary process for me and it will probably continue to evolve as long as I am able to spend time in my shop.

I am not disparaging small machines here but if your largest machine is a Grizzly mini lathe which weighs in at under 100 pounds then you may not need a material handling system like those being discussed

A drop of cutting oil on the tip of a self tapping screw will work wonders. After all, it is first a tool and then a fastener.

S. Roller ramps made for trailer. With these ramps, a couple of 2 X 8 planks and the hand cranked winch at the front end of the 16' bed I can load most anything within the hauling capacity of the trailer.

here. But whatever you do in your shop I bet there are times when you can use some sort of a device for making heavy parts easier to handle.

Other ways of doing it are plentiful. The hydraulic engine hoists used by mechanics work well in some environments and they are not complicated to build if you want to take the time. But the engine hoist always seems to me to be in the way when you are not using it. Fork trucks and pallet jacks have their place. You will have to decide what is the most useful for your application and you will probably be the one to build it. Remember the rules I mentioned at the beginning of this chapter and also remember this: whatever type of system you build — I didn't tell you how to build it!

Chapter 6

Repairing a Worn or Damaged Shaft

Rebuilding used machinery can be a satisfying experience and, if you know just how far to go in the process, it can be profitable. It is possible to invest more time and material in the rebuilding of a machine than the machine is worth. I am something of an expert on that subject, but we won't go into that right now. There are ways of minimizing the costs of these projects and what I hope to demonstrate with these photos and paragraphs represents some choices when a worn, bent or otherwise damaged shaft is the problem.

I should point out here that when this article was originally published in the Oct./Nov., 2004, issue of *Machinist's Workshop*, I received some mail about it. Some of it was good and some of it was bad but a particular comment by a machinist whose opinion I respect was critical of the stubbing process of shaft repair. He said, and rightfully so, that the example given here in this chapter would result in a somewhat weakened shaft.

There are repairs and there are repairs. Some times there is no really good way to permanently make repairs without replacing the damaged part. All of the methods discussed here can result in a somewhat weakened condition of the original part, building up with welding being the possible exception to this. You should, when using any of these processes, evaluate the service expected of the repair and, in many cases, make these repairs only as a temporary measure intended to serve until a new part can be made or purchased. Common sense and experience (there is that word EXPERIENCE again) will dictate to you which direction to take in shaft repair as well as any other repair process.

There can be many causes for a shaft becoming too small for the bearing, coupling, pulley or other accessory that resides on it.

Owning a lathe and a micrometer or a welding machine does not make you a metal worker any more than owning a piano makes you a musician.

It is not unheard of that the shaft may have been too small when the machine was originally built. Lack of lubrication in a bearing can also contribute to the condition. Most of the time, however, hours and hours of use doing just what the machine was designed to do will cause the shaft to wear and, as the condition worsens, the rate of deterioration can accelerate. The result can be a mechanical bearing turning on the shaft, a bronze or babbitt bearing becoming damaged, or "wiped," a wallowed-out keyway or any of several other symptoms of mechanical failure and it goes downhill from there until there is a total failure of the shaft and the machine. Not to mention the results of bad luck (none of us are careless, of course!) when a motor armature rolls off the work bench and gets bent or some similar disaster visits us.

Here are four different approaches to some of these problems. There are others but some of the other processes require more in the way of specialized machinery or equipment. These four basic processes can be done in a shop equipped with the tools most commonly found in a small machine shop.

PROCESS NUMBER 1: Split Bushing

Shown here in Photo number 1 is the worn area on the lower shaft from a 36" wood cutting bandsaw I rebuilt in my shop for a customer. You can clearly see the worn area where the bearing was removed. The bearing in this case was a 45mm double row ball bearing which has an inside diameter of 1.7717" and the shaft measured between .005" and .008" undersize. Even with the products available today for mounting bearings on undersize shafts it was clear that this shaft would not serve and must either be repaired or replaced. If the worn area is at the end of a shaft it is possible to build up the worn surface by welding and machining back to the required diameter. We will take that up later on. But when the shaft extends beyond where the bearing runs, as in this case, then you run the risk of the shaft becoming distorted and you are no better off than when you started.

1. Worn area on shaft.

I began this job by first machining a sleeve which will be used to repair the worn area. The sleeve, shown in Photo number 2, is machined to an outside diameter of approximately .050" larger than the finished diameter will be. The length is important in this case as we do not want the repair to extend into the threaded part of the shaft. Machine the inside diameter of the sleeve to a diameter of about 3/8"smaller than the finished repair will be. You will have to make some judgments here. If the shaft has a high lateral loading, for example a long extension with a pulley on the end of it, then you will not want to weaken it unnecessarily by cutting too deep. On the other hand you must have enough material to absorb the heat of welding. For a shaft of this size I have found that a wall thickness of approximately 3/16" works best. On bigger diameters I will go up to 1/4" wall thickness but there is no advantage to making it any thicker than 1/4". I chose to make the sleeve before machining the shaft for this reason. If I am going to machine two parts to fit together, an inside diameter and an outside diameter, and I have a choice, I will machine the inside diameter first. I find it sometimes saves time as I can machine an outside diameter to size more quickly than an inside

2. Sleeve machined to an outside diameter .050" larger than finished diameter.

3 & 4. Setup for machining worn area of shaft

diameter. Probably just me.

Photos 3 and 4 show the setup for machining away the worn area of the shaft. There is no need for indicating the shaft in or machining between centers at this stage. I mounted the part in a three-jaw chuck and machined the area where the sleeve is to go as close to the chuck as possible for rigidity. I used a carbide parting tool to machine the groove to the exact width of the sleeve and to a diameter to match the inside of the sleeve. Next I cut the sleeve in half and ground a weld prep on both sides of both halves.

Photo 5 shows the shaft with one half of the sleeve in place and the other half ready to go into place. The stainless steel hose clamp is put on the threads to protect them from weld spatter.

In Photo 6 you see the inserts in place and clamped. Clamp them tightly to the shaft before welding as they will shrink to the shaft after welding and make a firm, tight bond. A word of caution here. Be sure of your alloy when making the sleeve. The alloy of the shaft is immaterial, but you should use mild steel for the sleeve so as to insure there is no hardening of the welded parts in the heat affected zone (HAZ). I welded this with E7018 and I used an electrode diameter of 1/8" so as to be able to completely fill the groove in one pass. Photo 7 shows the welding in process and Photo 8 shows the sleeve welded in place. The area in Photo 8 where it looks as if the threads have been welded is where the weld ran onto the hose clamp and is not welded to the threaded area.

Set the shaft back up in the lathe and this time make sure it is running true. If you have a good three-jaw chuck, use that, but otherwise set up in a four-jaw chuck or run the shaft between centers. Machine to finished diameter and compare it to Photo 9, the completed repair.

5. Shaft with half of sleeve in place.

6. Inserts in place and clamped.

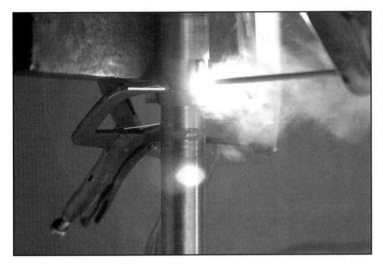

8. Sleeve welded in place.

7. Welding in progess.

9. Completed repair.

PROCESS NUMBER 2: Stubbing

I think the term "stubbing" must be a nautical term along with "Avast" and "Ahoy." I learned this in the Navy and I have never seen the term applied to a similar process anywhere besides there. But it is a good fix when a shaft, particularly a motor armature shaft, becomes bent or otherwise damaged beyond the point where it can be repaired with a thread file or emery paper.

10. Shaft with worn bearing surface.

The example shown here, see Photo 10, is another shaft with a worn bearing surface but here we will replace the entire shaft from this point out to the end instead of just replacing the worn area. Begin (this is important!) by taking careful measurements of the part you are going to repair.

Making a drawing of the job is a good idea. Then saw the bad end of the shaft off and set the good end up in a lathe.

If the shaft is large you will probably need to set up a steady rest but this will depend on the size of your lathe, particularly the size of the hole through the spindle. Face the end to get a clean, square surface. Drill and tap a hole for the stub. Here again you will have to make some decisions based on the size of the part you are working on. This example is a 1-1/2" diameter shaft and I drilled and tapped for a 5/8"- 18 stub. Obviously, you cannot

11. Two parts ready to be screwed together.

12. Drilling hole in shaft.

use a 14"-24 thread here nor would you want to use a 1-1/4" thread. Choose a thread size large enough to do the job yet still leave adequate material in the outer part. Common sense usually will dictate the proportions here. The two parts are shown ready to be screwed together in Photo 11.

Screw the stub into the end of the shaft and tighten it securely with a pipe wrench. You don't plan on removing it so screw it down tight. Then drill a cross hole through the shaft where the stub is. In this example I drilled through with a No. 1

drill bit and reamed to accept a #5 taper pin.

The important thing here is not the size of the pin but the quality of the fit. Be sure to ream the taper all the way through so there is no room for movement to take place after assembly. Cut the ends of the pin off and file them smooth. See photos 12, 13 and 14. Now set the shaft back up in the lathe and machine it back to the original specifications.

There are a couple of refinements to this method. If you know the direction of rotation of the shaft when it is in service you may want to use left hand threads instead of right hand. Another option is to use a spot of weld instead of a taper pin to keep it from unscrewing itself.

The method shown here, if properly done, will make a repair which can last as long as a replacement part.

14. Smoothed end of pin.

It doesn't make any difference what kind of a high-dollar, fancy, multi-axis positioning table you put on your drill press — it is still a drill press. And it still has a quill and a spindle which is designed for drilling and not for milling.

13. Fitting taper pin.

PROCESS NUMBER 3: Build up by welding

This option is a good one where the worn or damaged portion of the shaft is at one end of the part. It can sometimes be used in other situations but you will always run the risk of heat distortion when you weld in the middle of a part. You should also be fairly sure of the alloy of the material you are working with. Some high-carbon steels will harden in the HAZ and will require that you anneal it after welding. If this is the case you may want to choose a different repair process.

15. Runoff sleeve tacked into place.

The part used here as a demonstration is a particularly good example of repairing shafts when possible rather than replacing them. This is the blade holding arbor of a large wood planer. It is apparent that replacing this part, either by the repair shop making a new one or by ordering a new arbor from the manufacturer would be expensive. And, in this case, the machine being repaired is so old that replacement parts are no longer available.

You can see from the pictures that the surface where the bearing resides is not a 100% surface. The grooves machined for the blades extend into the area of the bearing mount. This was probably done when the machine was built to minimize the overall length of the assembly. It does not affect the repair method.

Begin by machining the bearing surface to about .020" undersize. You should not machine so small as to require more than one pass of weld metal deposit yet it is good practice to have clean, new metal for the welding. If the wear is extensive then you may have to use more than one pass. In most cases, however, you are just going to need one pass.

Nobody will ever take quite as good care of machines and tools as the person whose money paid for them.

Begin by making a bushing or sleeve, shown here tacked into place, in Photo15, to use as a place to start or end your arc. When welding plate this is called a runoff tab. It doesn't need to be a close fit onto the shaft but the OD should be the same as the part you are going to weld. Be sure when welding to avoid creating slag pockets (I know — I never leave any either.)

Weld all the way around the part with a heavy enough deposit of material to insure that you have more than enough material in place to clean up when machining. See Photos 16 and 17.

16. View of weld with heavy deposit.

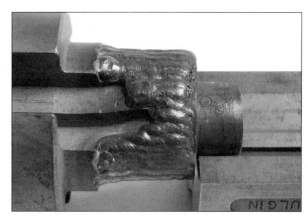

17. Welding completed.

18. Finished job.

After allowing the part to cool slowly, preferably wrapped in an insulating material, return the part to the lathe and machine the new bearing surface, machining away the "runoff bushing" in the process. I used a small grinder here to dress up the edges of the grooves where the blades will go. The inished job is shown in Photo 18. This job was a money maker for me and a money saver for my customer and that is what makes the world go around.

PROCESS 4: Shrinking on a sleeve

The final method we discuss here is like the first three in this regard: it has a definite place and can be a useful method for reclaiming worn parts but it isn't used everywhere. I have used it back in the days before I became a perfect machinist (AHEM) to recover from the mistake of turning a smaller diameter back past the place where it should have ended. Of course, I never do things of that nature anymore.

19. Corroded area needing repair.

The part selected to demonstrate this process is an axle out of a 3/4 ton International truck which was being restored. Photo 19 shows the corroded area where the grease seal runs on the axle and the reason for the required repair. None of the other processes discussed here will work. You obviously don't want to weaken the axle by using Process 1. You can't stub it. Welding would create some undesirable stresses. And replacing the axle would require a major junk yard search. So we will do it this way.

Set the axle up in the lathe and be sure it is running true. Machine the corroded area off and machine down to within approximately .025" of the bearing surface. Remove the axle from the lathe and make a sleeve of the proper length and about .050" larger than the original diameter. There will be some differing of opinions about the inside diameter but this is the way I do it and it seems to work. I machine a sleeve smaller than the diameter it goes on using the following formula: .010" plus .001" for each inch or fraction of an inch of diameter. For this job, 1.580" diameter, I made the sleeve to provide .012" interference. Photos 20 and 21 show the part being machined to size and the sleeve ready for installation. Note the chamfer in the sleeve to insure that it will seat solidly against the shoulder.

20. Part being machined to size.

21. Sleeve ready for installation.

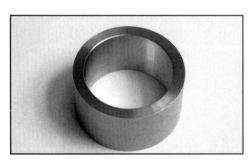

Set a pair of inside calipers, Photo 22, to a diameter of .002" to .003" larger than the diameter of the axle. This will allow you to measure the part as it expands while being heated.

The process of installing the sleeve is shown in Photos 23 and 24. Photo 25 was taken just before the part settled into place at the

I wonder just how many steer hides it would take to make a suit that would protect you while air-arcing overhead.

22. Taking inside measurement.

shoulder on the axle. Have all your tools at hand, and when you start to place the heated sleeve on the shaft don't waste any motion! When the sleeve first contacts the cold axle it is going to start to cool, and therefore to shrink, immediately. Be prepared to drive it into place or you will find yourself machining off the first attempt and waiting for things to cool down before you try again. Finish the job off after the parts cool down by going back to the lathe and machining the sleeve to the required dimensions.

23. Process of installing sleeve.

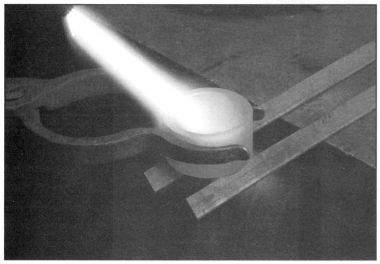

24. Installing sleeve.

25. Taken just before part settled into place at shoulder on axle.

26. Finished project.

Photo 26 shows the completed repair job.

As I said at the beginning, there are other methods of repairing shafts. The metal spraying process is a good process but it usually requires an investment greater than the small shop can afford unless you have a large volume of this type of work. In textile mills, rolling mills, and other manufacturing centers where rotating machinery is a maintenance issue, a good metal spraying outfit will quickly pay for itself.

There is also the option, sometimes the only option, of making a replacement part. Where the shaft involved is nothing but a piece of 1018 or 1045 cut to length with maybe a couple of keyways cut in it repairing the original doesn't make much sense. But for the small job shop one or another of the methods discussed here can make you some money and will assuredly add to your reputation of being "the guy who can fix anything!"

I like that part!

Chapter 7

Building an Ornamental
Fireplace Screen

I keep my roll-around toolbox close to the machine where I am working, but I keep all the drawers closed. There are few things more useless than a toolbox drawer full of chips.

The opinions expressed by the work force in my shop may not directly reflect the opinions of the management. If you want to know what management's opinion is you will have to go ask her.

This is a project which will, or at least it can possibly, do a number of things for you. If you make this fire screen and install it in your fireplace at some time when your wife is not at home, and she comes home and finds it there, you can possibly come off as the greatest hero since Superman arrived in his comet. Another thing it can do for you is make you some money if you have some way of getting to the people who would be happy to spend money for an item of this nature for their home. And, believe me, there are many people in that category. And the last, and possibly the most important, thing it can do for you is give you the satisfaction of knowing that you have the ability to make something as useful, as ornamental, and as downright practical as this fireplace screen is.

From 1973 through 1979 I built, along with my father and three other craftsmen who worked at my shop at the time, more than 2,000 of these fireplace enclosures along with thousands of shovels, pokers, brooms, tongs and andirons that go along with them. We sold a lot of them locally to individuals who heard of us through word of mouth, but primarily we sold them through a chain of fireplace accessory stores known as King's Row Fireplace Shops. We didn't do too badly at it until the energy crunch of the late seventies came along and the consumers suddenly all wanted fireplace enclosures with glass doors in order to save on heating costs for their homes. We made several of the glass door versions. I include a couple of photos of them here, but we ran into a real problem obtaining the tempered glass so we finally gave it up. I still get the opportunity from time to time to build one of these screens, both with glass and without, and it still gives me some satisfaction when I do it. Not to mention the fact that I charge a lot more for one now than I did in 1977!

For those of you who might be interested, there is a bit of history in the development of this particular design of fireplace enclosure. My father started making screens for customers sometime during the 1950s. He first built a free-standing screen with lifting handles on it but the development of high-dollar homes here in these mountains was catching on by that time and fireplaces just kept getting bigger and bigger. The free-standing units, if built big enough to provide protection from sparks and embers, were getting too heavy for the lady of the house to handle. So he came up with the idea of making gates and attaching them to the sides of the fireplace. The original gates had decorative hinges which were bolted directly to the masonry on either side of the fireplace. But it was a problem. If the anchors for the hinges were drilled into the masonry joints then there was a degree of precision required in taking the measurements for the screen and in attaching the hinges to the screen which was difficult to maintain. Drilling directly into the rock was really difficult. This was before the days of the readily available electric hammer drills which are popular today and drilling into a vertical surface was tough because you didn't have much way to bring pressure to bear on the drill point. Brick fireplaces weren't too bad but stone masonry was, and is still, popular in mountain homes and bolting a metal strap hinge to a rock wall and having both the screen doors line up and hang parallel was really difficult to do without taking so much time that the profit from doing the job just seemed to evaporate.

MATERIALS LIST
Ornamental Fireplace Screen

1. 2" x 2" x 3/16" angle iron — Enough to make 2 sides and the top of the main frame. Allow at least 6" more than the total of the three sides to provide coping at the corners.

2. 3/16" X 1-1/4" hot rolled strap — door frames — figure 3 times the width plus 4 times the height of the fireplace. Include a couple of feet extra if you decide to make and install the tool holder I made for this example.

3. 1/8" X 3/4" hot rolled strip — backing strip for door frames — takes the same length as item 2 plus an additional piece for the center part between the doors.

4. Plain steel woven wire cloth — .035" wire diameter X 6 openings per inch. — McMaster-Carr catalog number 9219T45. Don't scrimp on ordering enough to do the job.

5. 3/8" X 3/8" hot rolled square steel — About 2 feet should make the handles and the door stop.

6. 3/8" X 1-1/2" hot rolled steel — 2 pieces 3" long for anchoring brackets. If you include the crane you will need about 2 feet more. And if you include the crane you will need a piece of 3/4" round HRS to make the post.

7. 3/8" - 16 X 1-1/2" long square head set screws. 2 required for anchors.

8. 1/4"-18 flattened steel expanded metal — McMaster-Carr catalog number is 9302T27 for a 24" X 24" piece. Enough for most fireplaces.

9. 3/16" X 1" hot rolled strip — approximately 12"— latch and crane brackets.

10. 1/8" X 1/2" hot rolled strip — latching strips. 6" is plenty.

11. Black iron rivets — unless your fireplace is as big as a spare bedroom 50 will be enough — 3/16" X 5/8" — round head — McMaster-Carr catalog number is 97300A667 for a package of 140 — plain steel.

12. 1/4" Hot rolled round — 12" should do it for tool holder hooks.

13. (4) steel weld-on hinges — 2" X 2".

14. 1/2" square HRS — 60" — If you decided to make a shovel and a poker. You will also need a piece of 16 ga. sheet steel for a shovel blade.

15. Paint of your choice.

High speed drill bits under 1/2" in diameter will mutate. When you begin the process of putting them back into order in the drill index you will find three 25/64" bits and no 3/8" drill. This is something like the phenomenon of socks and the washing machine.

1, 2 & 3. Process of measuring a fireplace.

Enter a stroke of genius! This is where I credit my father with a flash of inspiration which resulted in both him and me making a pretty good living for a few years. In order to get away from the tedious job of drilling into masonry walls to install a fire screen, he came up with the one-piece design shown here in this chapter. It makes installation a breeze and it makes for a convenient and attractive way to provide protection from the coals that can pop out of the fireplace and possibly start a fire where you would prefer not to have a fire. The contrasting expanded metal panels at the bottom of the doors was the result of his not having quite enough screen material to complete a job so he put the expanded metal in and liked the results. The rest, as they say, is history.

So if you want to build one for yourself here is how it is done. This the way I have done it over the years and if you want to do it exactly as I will describe here, then go to it. But you will likely find different ways to accomplish the same thing. For example, the machine I use here to twist the screen handles and the poker and shovel handles is one I designed and built nearly 30 years ago and I doubt if you will want to go to the lengths required to build such a machine. But there are alternatives to twisting the steel and you may even come up with one nobody else has thought of yet. One alternative worth mentioning is a lathe in back gear or in a really low gear and a four-jaw chuck. We will go more into detail about twisting HRS sections in Chapter Eleven.

I am including a materials list with this chapter but keep in mind that it is a general list only. Your materials will depend upon the size of your fireplace and, to some degree, the methods you may employ to build the screen.

One further note: I forged the tool holder, the latch, the crane, and the shovel, poker and broom handles. If you do not have access to a forge you may choose to use alternate methods for all of these components. As I said before, I will describe how I went about this project. Any method you use that comes up with a product you are satisfied with is the correct way of doing it.

You will begin, obviously, by measuring the size of the fireplace opening. Several things must be considered here. The fireplace I am using here to demonstrate this job is an easy one. It is of rock but the sides and the hearth are very smooth and regular. Brick fireplaces are usually pretty easy to work with. Some fireplaces are really rough, I think the owners prefer the term rustic, but I think of them as just plain rough. All a matter of taste. An added advantage for me is that the fireplace in this example was built for my brother and I am older than he is so if I tell him to like it he has to like it. Try that with your wife!

Photographs 1, 2 and 3 show the process I use to measure a fireplace. Place a spirit level on each side in turn and note the high places or the places which protrude into the opening of the fireplace. Your width measurement should be taken at the point where the width of the opening is the smallest. To help to clarify that think of it this way. If a stone protrudes 1/4" into the opening near the top of the right side and another one protrudes 1/4" into the opening near the bottom on the left side your measurement should be 1/2" smaller than the nominal size. Measure the height from the hearth to the bottom of the steel lintel and pay attention here to the condition of the hearth. If the hearth has high places in it you will have to take this into consideration when making the doors.

4. Small parts needed in final assembly.

Experience, sometimes expensive experience, has taught me this. When I am building a fire screen for which I myself have taken the measurements I build to the nominal size. If anyone else measured the fireplace opening I leave 1/4" on each side and 1/4" on the top for clearance. No exceptions! This doesn't mean that I don't trust anyone else's ability to use a measuring tape. It merely means that when I am going to pay the price for a simple error I prefer to be the one who makes the error.

Now to the shop. I begin a fire screen by first making all of the small parts which will be required in the final assembly. This includes the latch, the handles, and all of the hardware and brackets necessary to complete the job. The photographs (in Series A) show the forging process for the latch and the tool holder. For the latch draw a piece of 3/16" X 1" HRS down so that you have a 3/16" square section at the end approximately 4" long. Twist it with the tongs, leaving 3/4" at the end untwisted, then bend it to its final shape. For the tool holder draw the end of a piece of 3/16" X 1-1/4" HRS out to leave a 1/4" round section at the end approximately 4" to 5" long. Form the circle or any other finial you may prefer at the end. The 1/4" round pieces for the hooks on the tool holder may be formed now while you have the fire in the forge or they may be formed cold. It is easier to do it hot, by the way. The handles for the doors will be formed cold. The crane is an option you may or may not want to include but if you do include it forge a piece of 3/8" X 1-1/2"

5. Coping the ends of the top piece to provide an assembly.

SERIES A: Making small fireplace parts

A1. Drawing out the latch

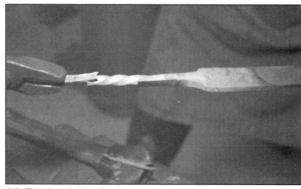

A2. Twisting the latch

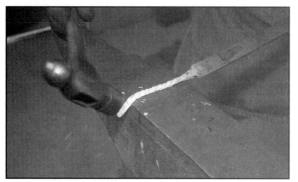

A3. Forming the latch

A4. Forming the latch

A5. Forming the latch

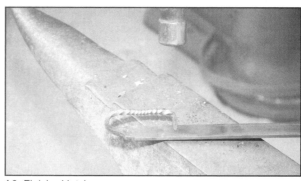

A6. Finished latch.

A7. Forming the tool holder.

A8. Forming the tool holder.

A9. Finished tool holder

A10. Bending the handles on the Hossfeld

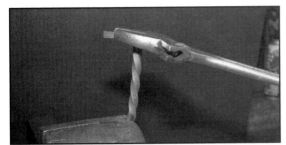

A11. Twisting the handles with an adjustable wrench

HRS to make a tapered part with a turned up hook on the end. Make the taper so that the top of the crane will be parallel with the hearth. The hook is a modified "S" hook as shown. All of the forged and miscellaneous parts, with the exception of the crane and hook, are shown in Photo 4.

Begin the frame by sawing the 2" X 2" X 3/16" angle iron to length. Saw the top piece to equal the fireplace opening width plus 3-5/8" and saw the two sides to the opening height minus 3/16." Cope the ends of the top piece to provide an assembly as shown in Photo 5.

If you do much welding you will know this but I will say it here anyway. When you clamp the angle iron parts to the table for welding, leave the angle a little greater than 90 degrees. There will be some distortion from the welding. If you anticipate this you can clamp the parts down with the legs spread at the bottom a total of about 3/8". Photo 6 shows the approximate amount of spread you should leave when assembling the frame. When you weld the corner, and then weld the anchoring bracket in, the distortion will draw the parts almost exactly square. And if

6. Approximate spread to leave when assembling frame.

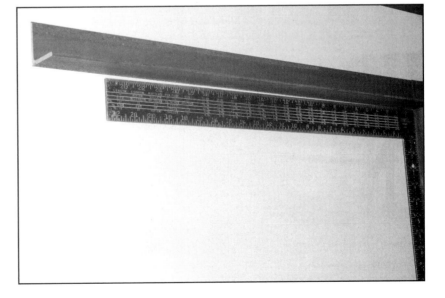

7. Door parts fitted into main frame.

8 & 9. Placement of anchoring brackets for welding and door step.

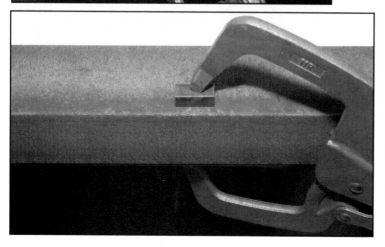

they are still out some it is an easy matter to bring them into alignment by striking the weld with a ball peen hammer. It is much easier to close up a corner of this type than to open it.

Now is the time, before you start welding, to take the measurements and saw the parts for the door frames. Photo 7 shows these parts fitted into place inside the main frame. (I hope you don't think I am going to do much welding on that pretty white background. I was just trying to make the photos more clear.) The important thing to remember to do here is to leave a gap between the two doors of from 1/8" to 3/16". This is also your opportunity to compensate for an irregular hearth. If the hearth is rough leave enough of a gap at the bottom of the doors to clear the highest point on the hearth. I usually will allow about 3/16" gap at the bottom regardless of the condition of the hearth. If the hearth is so rough and uneven that the gap required to clear it is too great you may need to weld in a horizontal member at the bottom of the screen and fit the doors to it. I have used either 1" X 1" X 1/8" angle or 1" square tubing for this purpose. After sawing the door frame parts lay them aside and weld the main frame at the corners. Do not weld the insides of the corners yet.

Turn the frame up and clamp the top of the frame to the table and clamp into place the two anchoring brackets. They should already be drilled and tapped for 3/8"-16 before welding them on. While the frame is in this position you may also weld the stop tab in the center of the frame. When you weld the anchoring brackets and the door stop in place, set them back from the front edge of the frame by 3/16" to allow for the thickness of the door frames. See Photos 8 and 9.

Then finish welding the frame at the corners and grind the welds smooth. If you are going to install a crane now is the time to weld in the crane brackets. We will talk about that at the end of this discussion when we will take up the optional features you may or may not want to consider. The finished frame is seen in Photo 10.

Begin construction of the doors by

clamping into place the six pieces of one door. See Photo 11. Notice that the smaller section at the bottom of the doors is proportional to the height. This is hard to describe but I leave an opening of from 4" up to about 5-1/2" depending on the size of the fireplace. The screen we are building here has basic dimensions of 41" X 28" and I left the opening at 4-3/4." Another part which should be welded in at this time is the tab for the latch. Weld it to what will be the right hand door at about the center of the height of the fireplace. Weld all joints, including the latch tab, on both sides and grind smooth.

The secondary door frames are made from 1/8" X 3/4" strap and are tacked to the main door frames as they are made. This makes it easier to build the frames for one thing, but the primary advantage is it keeps the parts in alignment while drilling for the rivets. Weld all the joints on both doors and grind smooth.

Using a piece of 3/8" material for a straight edge, scribe a line down the

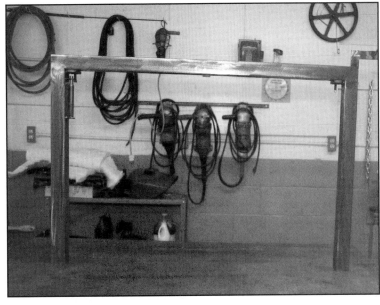

10. Finished frame.

11. The six pieces of one door clamped into place.

center of all sides of the secondary frames of both doors. Center punch at each intersection at the corners and then space the rivet locations along each side of the frames using a pair of dividers. It is worth the time it takes to have the rivets spaced equally around the periphery of the frames just because this sort of attention to detail will add to the quality of your work. At least, that is my opinion.

12. Relationship of backing frame to door frames. and method of marking rivet holes.

Center punch all of the marked holes and drill clearance holes for the 3/16" rivets. In Photo 12 you can see the relationship of the backing frames to the door frames as well as the method for marking for the rivet holes.

13 & 14. Before and after de-burring of corner.

After drilling all the holes, match mark the frames with a center punch or a number stamp so that when you take the backing frames off you

can assemble them in the proper order. I don't care how careful you are when drilling the holes there will be enough difference in them so that some of the rivets will not go in place if you mix up the frames. Cut the tacks holding the secondary frames with a cold chisel and finish all of the grinding, filing and de-burring of all the door parts at this time. Photos 13 and 14 show corners before and after de-burring. Just one more indication of the attention to detail that will make the difference between an adequate job and a really good job. Not everybody who sees the finished product will be aware of this degree of finishing, but I would be. And if you don't take the time to do it, you will be too. It is called craftsmanship.

Now we are ready to assemble the doors. Cut the wire so that it comes as close to the edge of the secondary frames as possible without showing on the outside. Clamp the parts together and, using a pointed awl which can be made from a worn out screwdriver, open the hole so that a rivet can be inserted. Do the

15. Dimple drilled in corner of welding table for upsetting rivets.

corners first. When you get to the bottom where the expanded metal goes you will have to use a drill to open the holes because the expanded metal will not allow the insertion of the rivets otherwise. It is better to place all of the rivets in their holes and upset them just enough to keep them in place before starting the process of hammering them down.

Photo 15 shows the dimple I have drilled in the corner of my welding table for upsetting the rivets. I built a machine which used an air hammer years ago when fire screens was a way of life around here but it is long gone now. Rivet the latch on after the screen is done making sure that the rivet has enough slack for the latch to move.

16. Piece of sheet metal used to protect screen from welding spatter.

Making the handles and the catches is largely a matter of choice and equipment availability. I bent the handles as shown on a Hossfeld® No. 2 ironworker and twisted them on the machine I mentioned at the beginning of this chapter. The handles are made from 3/8" square HRS and are made here without heat. If you have a forge, or a torch,

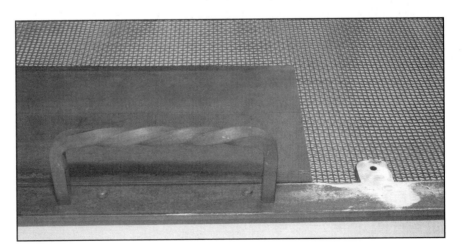

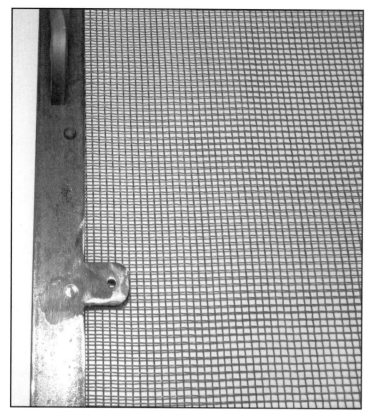

it may be easier for you to heat them to bend them. The 3/8"-square HRS can be twisted cold with just an adjustable wrench as may be seen in one of the photographs or you may prefer not to twist them at all. It is largely a matter of choice.

Weld all of the remaining parts to the doors before assembling them to the main frame.

Photos 16 through 19 show various views of the process. Notice in Photo 16 the piece of sheet metal I use to protect the screen from welding spatter or, more importantly, to protect from an accidental arc strike on the screen. That can really mess up your day! And in Photo 17 you can see that the rivet which lies under the latch has been ground flat. It is a good idea to countersink the hole where that rivet goes so that grinding away the head will not cause the rivet to come out.

17. Rivet under latch has been ground flat.

It is much easier to get all of the welding and grinding done on the doors as sub-assemblies than to try to do it after the screen is completely welded up.

Photo 18 and Photo 19 are of the catches for the hinges and the latch being positioned for welding. I do not show a photo of it but you will also want to weld a piece of 1/8" X 3/4" strip down the back side of the door opposite the latch to cover the gap between the doors.

18. Hinges being positioned for welding.

We are almost done! Place the main frame back on the welding table and clamp the hinges to it at the appropriate points. Don't forget to put your sheet metal scraps back in place next to the hinges. Weld the hinges on all sides and grind the welds smooth. Photo 20 is of the completed job before painting.

I mentioned earlier that there are some optional features to be considered.

These include tool holders, cranes and the shovel and poker shown here. The crane was an essential part of the fireplace in the earlier days when the fireplace was used for cooking. Its use is somewhat limited now, but you never know when the power is going to go out and you will have to at least boil water for coffee in your fireplace. And when I was selling these things I told potential customers that the crane makes a good place to hang a decorative pot of flowers in the summer. I sold a lot of cranes that way!

19. Catches for latch being positioned for welding.

The crane brackets, Photo 21, are simply two pieces of 3/16" X 1" strap with 9/16" holes drilled through them. The crane post is turned from a piece of 3/4" round HRS and is made so that it can be installed by pushing the top up through the top hole and then let the bottom journal rest on the bottom bracket. The finished crane may be seen in the photograph of a screen which was at one time used here as a demonstration model. The unique thing about this screen is the set of fire tools which are hanging on it. My grandfather made the tongs for this set. My father made the shovel. I made the poker and my son made the handle for the broom. Four generations of workers represented in one set of tools.

The tool holder is made from a piece of 3/16" X 1-1/4" strap. Turn the eye on the end and cut to length. Some fireplaces are small enough

20. Completed job before painting.

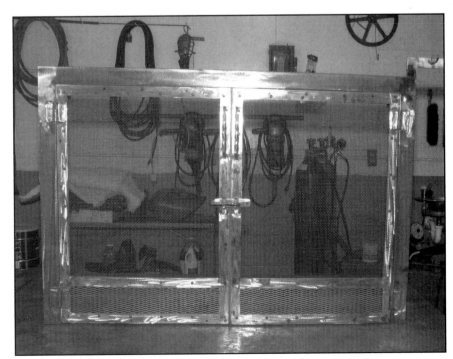

21. Crane brackets.

We now have cordless telephones, cordless drills, cordless saws, cordless just about everything. I can't wait for the cordless 300-Amp welding machine.

that you will want to offset the tool holder as shown in one of the photographs of fire screens included here. Tool holders can have as many or as few hooks as you desire and may go on either side, or both sides, of the fireplace. This is another good place for you to incorporate your own ideas. If you already own a set of fireplace tools look at the handles on them to make sure that they can hang on a simple hook. If not you may have to modify either the handles or the tool holder hooks or even use a different method for the storage of your tools.

Making the shovel and the poker is another fairly simple exercise in forging and welding. If you are familiar with the "Foxfire" series of books produced by the Rabun Gap Nacoochee School in Rabun County, Ga., you may have seen this same poker made by my father some years ago. There is a discussion of several blacksmiths and their techniques in *Foxfire Five*. We will talk about various methods for twisting steel shapes in Chapter Eleven so I will not repeat it here.

The finish chosen here is a favorite of mine. I sprayed the screen with a clear polyurethane finish. I like the bare metal appearance but be aware that this finish will not hide welding booboos like black paint will. Of course, I know that you have not made any welding booboos in this job. I experimented with several types of paint when I was in this business. When I had to paint six or eight screens a day I used an automotive radiator paint and a Binks® No. 7 spray gun. There are some high performance paints available which will withstand higher

Series B: Examples of Ornamental Screens

B1. A good idea here of what the crane looks like in place. Four generations of metal workers are represented in the tools shown here.

B2. A screen made to fit a full arch fireplace.

B3. One of the early glass door fireplace enclosures. Notice the damper at the bottom.

B4. A fireplace curved in the other plane. Like the side of a silo.

B5. An unusual shape but the screen turned out nice.

B6. The most difficult fireplace I ever fitted. Glass doors, too!

B7. A close up of the mark I put on all of my ornamental iron work.

More famous last words. I am only going to drill one hole. No need to clamp it down just for that.

temperatures but they were never worth to me what they cost. I found that the paint from a spray can of flat black enamel would last almost as long as the more costly paints and was much more readily available. It is not a problem to take the screen out in the spring and repaint it.

Installation is a piece of cake. Simply set the screen in place and tighten up the setscrews on either side. If you have measured correctly and allowed for all of the inequalities of the masonry there should be no problem. When I went out to install screens I usually kept an assortment of shims in my toolbox to put under one or the other side of the frame. I probably have left $20.00 worth of pennies, nickels and dimes under the corners of screens over the years. Photo 22 is a picture of the completed job. All that is lacking is a cold winter evening and a cup of hot cider!

If you are proud of your work you should mark it in some way. I have a stamp with which I mark all the ornamental iron pieces I make. Over the years I have received mail, and some business, by the way, because of the marking I put on a job. There are with this project, as there are with all projects, many opportunities for you to make changes to fit your own material availability or your own preferences. The machinery available to some may not be available to all. Some of you will have power hammers and iron workers while others will have only a welding machine and a cutoff saw. But the ingenuity of the hobby metal worker is endless. And that is what will make the difference.

22. Finished product.

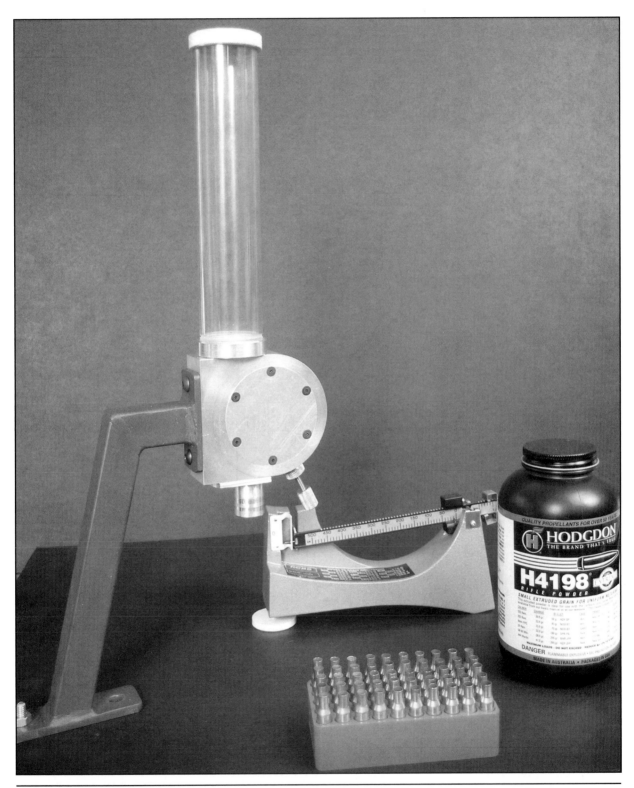

Chapter 8

Building a Better Powder Measure

T his machining project is made-to-order for the home shop machinist. It requires the use of the lathe, the milling machine and the band saw. It includes some interesting angles and some rotary table work. There is a little bit of welding and it even calls, as an option, for the use of a vertical shaping attachment if you have one. It isn't terribly difficult but it offers some challenges. And best of all, it provides us with the opportunity to spend maybe $400 worth of time and material to produce an item which can be bought where you purchase your reloading supplies for about forty bucks. What more can be asked of a machining project? And this one has the bonus of actually being a useful item which, I think, is in some ways superior to the ones now offered for sale.

This is another project of which I now own two examples. I made this measure to do some reloading for myself, and when *Machinist's Workshop* indicated it might make a good construction article, I had to build a second one in order to take the required photographs. The article appeared in the June/July 2006 issue of *Machinist's Workshop* magazine.

Bolt, rivet, nail, GTAW, GMAW, SMAW, SAW, silver solder, soft solder, braze or glue? That is the question. Knowing when to use which process for joining materials together is as important as knowing how to do it.

A couple of important notes here. If you are an experienced reloader of ammunition you will know that a powder measuring/dispensing device like the one described here dispenses according to volume. You will need to set it, and check it frequently with an accurate scale, for the particular ammunition you are loading. A metering device made to the exact dimensions shown in these photos and drawings has the capacity to measure up to .147 cubic inches (2.4cc) so if you reload for some of the hand cannons available today you will have to scale up the project. The materials list included here is for the dispenser I made. As is always the case, there are many opportunities for substitution in both size of components and the material choice but this worked out pretty well for me.

We begin the job by laying out the 1-1/4" aluminum plate as shown in Photo 1. Exact sizes are given in the drawings but the outside dimensions are really not important. Nor is the outside shape. Just as long as it is big enough to accommodate the inner parts and it is pleasing to you. Locate a center point and scribe the circle and the 2 tangential lines. I sawed the rough profile out on my vertical bandsaw. Photo 2 shows a good method for centering the roughed-out part in a 4-jaw chuck. Position a parallel to keep the faces of the work piece running true while using the live center to hold it in place. While holding a light pressure on the center, tighten the chuck jaws around the part. BE SURE TO REMOVE THE PARALLEL BEFORE YOU TURN THE SPINDLE ON! I know I didn't need to say that but sometimes it is good to be reminded. Bore to the desired dimensions and remove the part from the lathe. See Photo 3.

I have cast iron machinery, a welding table made from 2" thick steel plate and a concrete floor. A coffee cup will last me an average of about 25 days.

There are many methods for centering parts on a rotary table but the one I used here is one I particularly like. You will see in Photo 4 the parts required for quickly and accurately centering this part, or any part, on the rotary table. On the left are (2) shims, 1/8" thick, sawed out to a slightly smaller profile than the part we will be cutting. Protruding from the center of the rotary table is a centering plug machined from a No. 3

MATERIALS LIST
Powder Measure

1. Aluminum plate: 3-1/2" X 3-3/4" X 1-1/4" thick.

2. Aluminum bar: 2-1/2" diameter X 6" long. This will be enough to make the spool, the dispensing nozzle and the reservoir end caps.

3. Aluminum sheet: 1/8" thick — a piece big enough to make the cover plate — about a 3-3/4" diameter circle.

4. Polycarbonate tubing: 1-3/4" OD X 1-1/2" ID X 12" long. McMaster-Carr Catalog number 8585K17.

5. One (1) 7/16"-20 SS bolt at least 2" long. You can make this from SS stock. You will need just an inch or so of body diameter and a short length of threads.

6. Brass rod: 5/8" diameter X 1-1/2" long will be enough to make all of the brass parts of the metering components. 303 SS will also work well here.

7. One (1) piece of 316 SS threaded rod 3/16" - 32 X 3" long. McMaster-Carr Catalog number 93250A120.

8. Ten (10) #6-32 X 3/8" Socket FHCS.

9. One (1) #6-32 X 1/8" set screw

10. Two (2) 1/4"-20 X 3/4" Socket head button head screws.

11. 1" X 1" X 16 ga. square tubing and a couple of pieces of 1-1/4" X 1/4" steel strap. This material and the 1/4"-20 screws are for the mounting bracket if you choose to mount it the way I did. You may have other preferences.

I have no doubt about the effectiveness of SPC in our modern industries. It is a remarkable tool. But I still think that statistics can be misleading. If you are standing with one foot in a bucket of boiling water at 212 degrees F and the other frozen in a block of solid ice at minus 10 degrees F then, statistically speaking, you are pretty comfortable.

1. Laying out the aluminum plate.

2. Above, a good method for centering the rough-cut part.
3. Right, bore to desired dimensions.

Morse drill shank. My rotary table has a No. 3 Morse hole through the center. On the right is another centering plug made for this project (save this plug after completing this operation because you will need it again), and there is also a #10-32 screw and some washers for holding the plug in place. You can see how all this comes together in Photo 5. When you mount the work onto the table be sure that the flat side is parallel to the Y axis of the milling machine and the rotary table is set to zero degrees.

Photo 5 also shows the cut which finishes the flat side of the body. I finished this surface by removing about .020" of material with the end mill. Lock the rotary table against accidental rotation and climb cut the surface for the best finish. This process is a good way to finish machining all the way around the part as well, but I elected to use my vertical slotting attachment to produce the profile. Photo 6 is a picture of the tool I made for this operation. I welded a short piece of a HSS cutting tool to a piece of 5/8" CRS. Conventional wisdom says not to weld HSS but my wisdom isn't very conventional sometimes and by using a 3/32" diameter stainless steel welding electrode and about 75 amps of DCRP current it works pretty well. Photo 7 shows the cut being made around the part. Machine parallel to the X axis of the milling machine until you come to the centerline of the part. Lock the milling machine table in place at this point and machine around the radius by turning the rotary table through 180 degrees of

Listen to the machinist who has 30 years experience. But beware the machinist who has one year of experience 30 times over. Just being there a long time doesn't make one an expert.

4. Parts required for quickly and accurately centering part.

arc. Reclamp the rotary table, unclamp the milling machine table and continue the cut parallel to the X axis until you come to the end of the part. While you are machining this profile is a good time to call whoever you are responsible to for expenditures in your shop and tell them, "Come and look. Here is a good example of why we needed to purchase this slotting attachment." Public relations are important to the home shop machinist.

Without removing the part from the rotary table turn the table on its end so that the part is now vertical in the milling machine. Rotate the part 180 degrees which should bring the top of the arch to the top, positioning the base or flat edge of the part at the bottom and parallel to the milling machine table. Check the base with a dial indicator if you are not sure. The witness mark on the rotary table should now indicate 180 degrees. Now position the milling machine table so that the spindle is in line with the center of the part in both the X and Y axes. Drill a series of 3/8" diameter holes around the arched outer profile of the part at approximately 15 degree increments starting at 115 degrees and ending at 245 degrees (180 degrees +/- 65 degrees). After using the drill to rough out the slot, machine the finished slot with a 3/8" diameter end mill. Machine the entire length of the slot on center then offset by .037" either way to produce the full width of the slot. Photos 8 and 9 show the holes being drilled and the roughing pass while Photo 10 shows the completed slot. I will tell on myself here in the hope it will help you to avoid a problem.

Notice in Photos 9 and 10 the 1/2" machine parallels spacing the part out from the rotary table surface. I set up originally using the aluminum spacers shown in Photo 3. When I started to machine the slot I did not have sufficient room for the spindle to clear the rotary table surface so I had to set the part up a second time using the 1/2" parallels as spacers. And all this time you thought I never made any mistakes!

Photos 11 and 12 show what I refer to as the spool or the rotating center part of this project. It is shown finished in the lathe just prior to being parted off. This is a simple job of turning and the dimensions here are important only as they relate to the hole in the body. This needs to be a loose enough fit to allow free rotation yet not so loose as to allow powder granules to get between the parts. I found that from .003" to .004" clearance seems to be good here. Notice that the chamfers on the ends of the knurled surface are machined after the knurling is done. This will give crisper corners and make for an improved appearance. Move to the milling machine or drill press and drill the 25/64" diameter hole through the center line of the spool. Tap one end of the hole for 7/16"-20 TPI. The #6-32 setscrew is installed by first milling a counter bore with a 3/16" diameter end mill, then drill with a #36 drill and tap through into the 7/16" threads. See Photo 13. This setscrew is necessary here because we do not want to run the risk of the shaft accidentally turning during use which would change the volume of the powder dispensed.

You will remember when we centered up on the rotary table I said that there would be another use for the centering plug and here it is. Machining the

Where is it ever going to stop? Materials are getting harder and tougher. Cutting tools are getting more aggressive. Machines just keep getting faster and faster. Seems like the poor machinist is just trying to hang on!

1/8" thick cover plate presents the problem of holding a thin, flat disc in order to cut its outside diameter. Photos 14 and 15 demonstrate how this is done. Place a piece of material smaller than the diameter of the part you are making in the lathe chuck and face it true. Then using the centering plug and a live center, sandwich the part to be machined between the

5. Machining bottom flat.

part in the chuck and the plug. Take light cuts until you reach the desired dimension. I have used this method to cut as many as 10 parts at one time when all are to be machined to the same diameter. Once the part is machined to its finished diameter, go back to the milling machine and drill and tap the (6) holes for installing the cover plate. This can be done by one of at least three methods. It may be set up on the rotary table and the holes drilled and tapped at 60 degree intervals around the part. Or it may be set up in the milling machine vise, indicated for center and using either the DRO or dial indicators to locate the holes. Or thirdly, laying out the holes with a scribe and a protractor. If you choose the last method, be sure to match mark the two parts for ease in reassembly.

6. Tool made for this project.

The parts shown in the drawings as part numbers 04 and 06 are the two parts through which the powder passes into and out of the measuring chamber. The dispensing nozzle, part no. 04, is machined from a piece of 2-1/2" material and then sawed or milled to its final shape. The .2" hole through the center is drilled with a #7 drill and the tapered

7. Cut being made around the part.

8 & 9. Hole being drilled; roughing pass.

bores are then machined with a modified boring bar as shown in Photo 16. The .2" bore is smaller than a .22 caliber shell case and yet the taper will allow up to .45 caliber to be charged using the same nozzle. Part 06 also serves as the bottom cap for the reservoir. Both of these parts are fastened to the body with (2) #6-32 X 3/8" flat head socket head screws.

The knurled top cap, part no. 07, is machined from a piece of 2" aluminum and needs no further explanation except to say that I machined the threads for the top cap to be a slight bit looser than the

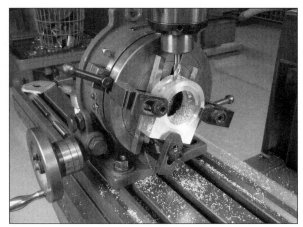

10. Completed slot.

bottom cap. When you remove the top cap to replenish the powder you do not want the reservoir to unscrew itself from the bottom.

11. Spool or rotating center part.

Machining the threads on the polycarbonate tubing, Photo 17, is easily accomplished but be careful in chucking the material. Too tight may deform or break it and too loose will allow it to move in the chuck while cutting the threads. Neither is good. I used my standard 3-jaw chuck with hard jaws but you may want here to machine a set of soft jaws. You will have to be the judge.

The six parts of the adjusting plug shown in the drawings as part details 09 through 14 and shown assembled in Photo 18 are fairly simple exercises

12. Cutting chamfer on knurled spool.

13. Installing setscrew.

14. Setup for machining cover plate.

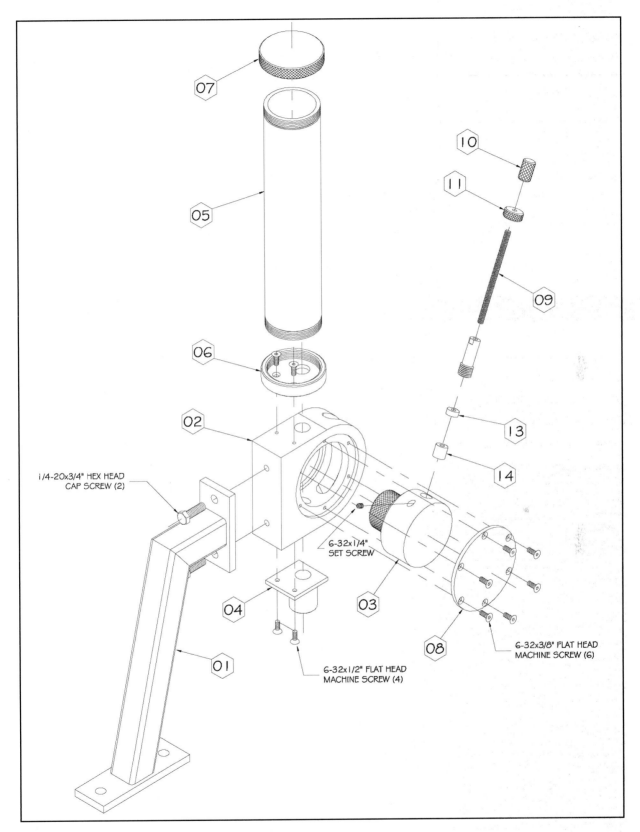

07

10

11

05

09

06

13

02

14

1/4-20x3/4" HEX HEAD
CAP SCREW (2)

6-32x1/4"
SET SCREW

03

04

01

08

6-32x3/8" FLAT HEAD
MACHINE SCREW (6)

6-32x1/2" FLAT HEAD
MACHINE SCREW (4)

Drawing No. 1

15. Machining cover plate.

I try to meet the needs of my customers. Now if we could just reach agreement on what constitutes an emergency.

in the lathe. Collets are the preferred method for set up here but a small 3-jaw chuck will work fine. I chose to purchase a piece of 10-32 stainless steel threaded rod (McMaster/Carr P/N 93250A120) instead of trying to thread the full length with a die. From left to right on the threaded stem are the plug which should be as close a fit in the spool as possible without binding, a locknut to insure it doesn't move on the shaft, the 7/16"-20 bolt through which the adjusting threads are tapped, a brass locknut, and finally the adjusting thumb screw. The thumb screw should be screwed on to the stem tightly, and this is a good place for a drop of LocTite®. The flats shown being machined on the threaded sleeve in Photo 19 are there to facilitate screwing it into the valve disc but are not absolutely necessary.

The mounting bracket can be seen in Photo 20 and is included in the drawings as part detail no. 01 but it is only a suggestion. Every one of you who build this device will have your own requirements and methods for mounting it near to or on your reloading bench.

Some notes about materials. I have chosen 6061-T6 aluminum for most of the parts because I tend to accumulate scraps of this material in my shop. And it is easily machined. I used stainless steel for all of the internal parts because there may be some corrosive effects from the

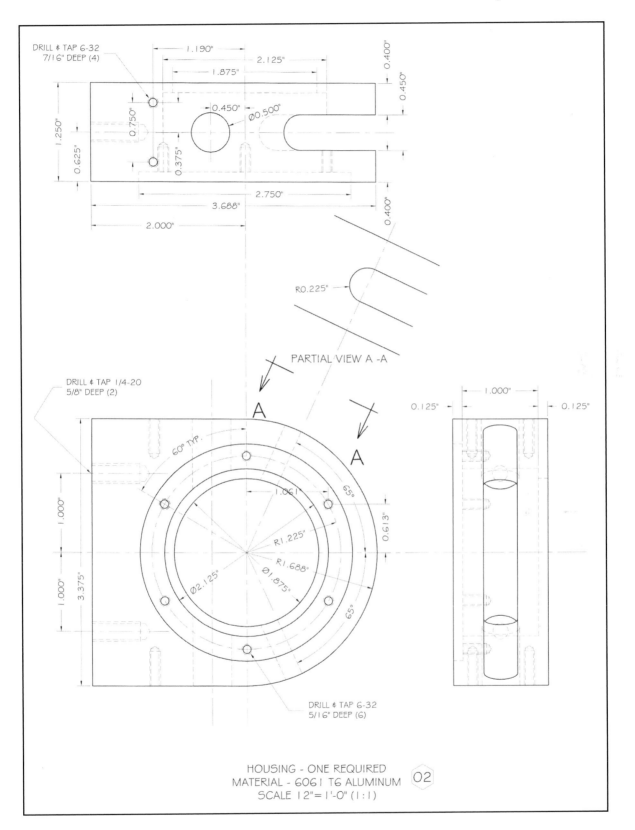

DRILL & TAP 6-32
7/16" DEEP (4)

1.190"
2.125"
1.875"
0.400"
0.450"

0.450"
Ø0.500"
0.750"
1.250"
0.625"
0.375"
2.750"
3.688"
2.000"
0.400"

R0.225"

PARTIAL VIEW A -A

DRILL & TAP 1/4-20
5/8" DEEP (2)

A

A

60° TYP.

65°

1.061

R1.225"

R1.688"

Ø2.125"

Ø1.875"

65°

1.000"

1.000"

3.375"

0.613"

1.000"

0.125"

0.125"

DRILL & TAP 6-32
5/16" DEEP (6)

HOUSING - ONE REQUIRED
MATERIAL - 6061 T6 ALUMINUM   02
SCALE 12"=1'-0" (1:1)

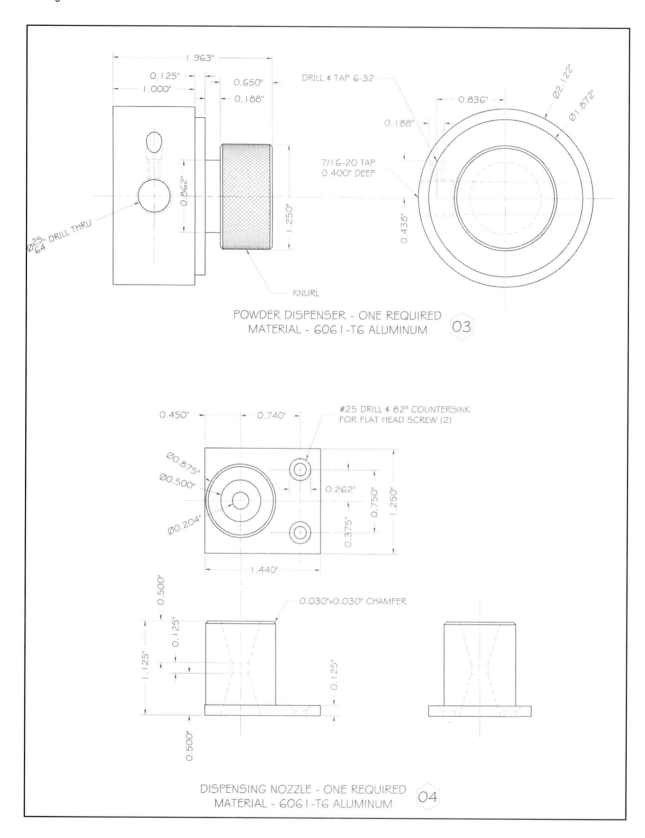

1.963"

0.125"
1.000"
0.650"
0.188"

DRILL & TAP 6-32

0.836"

Ø2.122"
Ø1.872"

0.188"

0.862"

1.250"

7/16-20 TAP
0.400" DEEP

0.438"

Ø 25/64 DRILL THRU

KNURL

POWDER DISPENSER - ONE REQUIRED
MATERIAL - 6061-T6 ALUMINUM   03

0.450"
0.740"

#25 DRILL & 82° COUNTERSINK
FOR FLAT HEAD SCREW (2)

Ø0.875"
Ø0.500"

0.262"

0.750"
1.250"

Ø0.204"

0.375"

1.440"

0.500"
0.125"
1.125"

0.030"x0.030" CHAMFER

0.125"

0.500"

DISPENSING NOZZLE - ONE REQUIRED
MATERIAL - 6061-T6 ALUMINUM   04

Drawing No. 3.

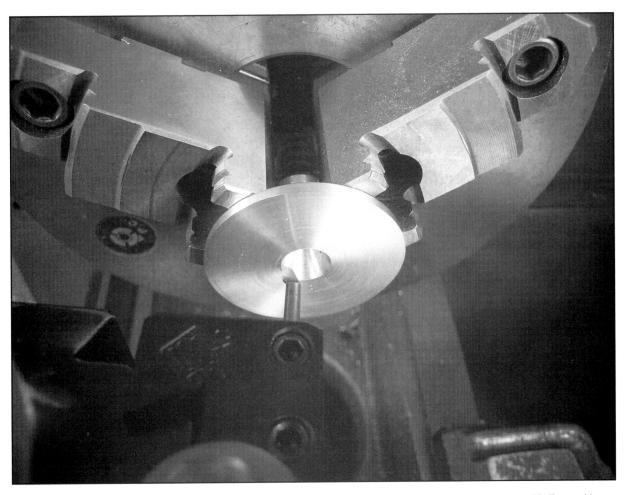

16. Tapered bores machined with a modified boring bar.

17. Machining threads on polycarbonate tubing.

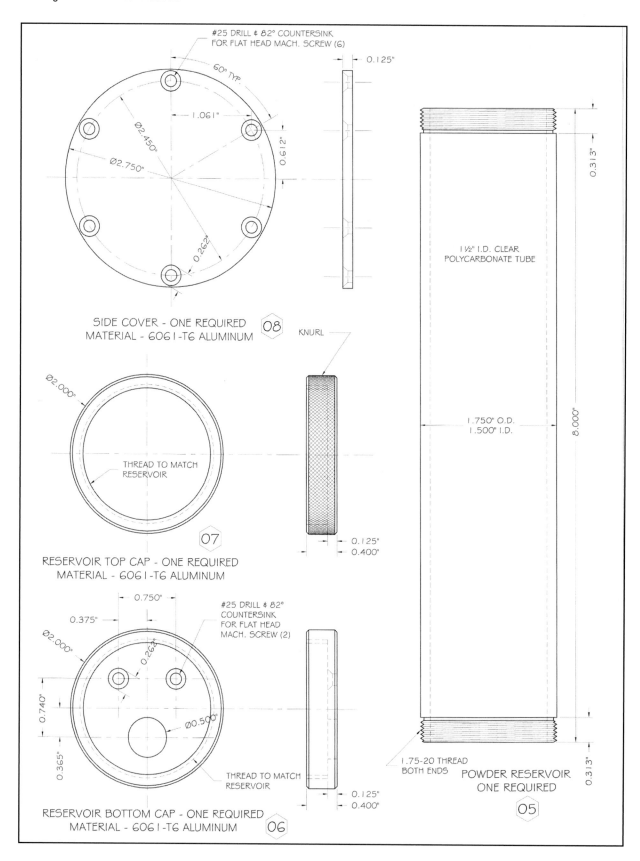

#25 DRILL & 82° COUNTERSINK
FOR FLAT HEAD MACH. SCREW (6)

60° TYP.

1.061"

Ø2.450°

Ø2.750"

0.612"

0.262"

0.125"

SIDE COVER - ONE REQUIRED
MATERIAL - 6061-T6 ALUMINUM   08

KNURL

Ø2.000°

THREAD TO MATCH
RESERVOIR

0.125"
0.400"

RESERVOIR TOP CAP - ONE REQUIRED
MATERIAL - 6061-T6 ALUMINUM   07

0.750"

0.375"

Ø2.000°

0.262"

0.740"

Ø0.500"

0.365"

#25 DRILL & 82°
COUNTERSINK
FOR FLAT HEAD
MACH. SCREW (2)

THREAD TO MATCH
RESERVOIR

0.125"
0.400"

RESERVOIR BOTTOM CAP - ONE REQUIRED
MATERIAL - 6061-T6 ALUMINUM   06

1 ½" I.D. CLEAR
POLYCARBONATE TUBE

0.313"

1.750" O.D.
1.500" I.D.

8.000"

1.75-20 THREAD
BOTH ENDS

POWDER RESERVOIR
ONE REQUIRED   05

0.313"

Drawing No. 4.

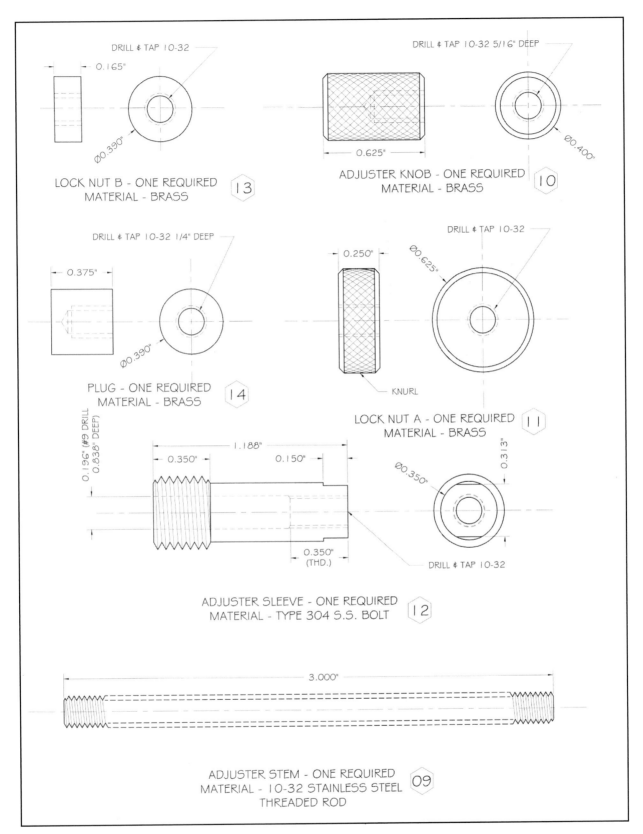

DRILL & TAP 10-32

0.165"

Ø0.390"

LOCK NUT B - ONE REQUIRED
MATERIAL - BRASS   13

DRILL & TAP 10-32 5/16" DEEP

0.625"

Ø0.400"

ADJUSTER KNOB - ONE REQUIRED
MATERIAL - BRASS   10

DRILL & TAP 10-32 1/4" DEEP

0.375"

Ø0.390"

PLUG - ONE REQUIRED
MATERIAL - BRASS   14

0.250"

Ø0.625"

DRILL & TAP 10-32

KNURL

LOCK NUT A - ONE REQUIRED
MATERIAL - BRASS   11

0.196" (#9 DRILL)
0.838" DEEP)

1.188"

0.350"

0.150"

0.350"
(THD.)

Ø0.350"

0.313"

DRILL & TAP 10-32

ADJUSTER SLEEVE - ONE REQUIRED
MATERIAL - TYPE 304 S.S. BOLT   12

3.000"

ADJUSTER STEM - ONE REQUIRED
MATERIAL - 10-32 STAINLESS STEEL
THREADED ROD   09

Drawing No. 5.

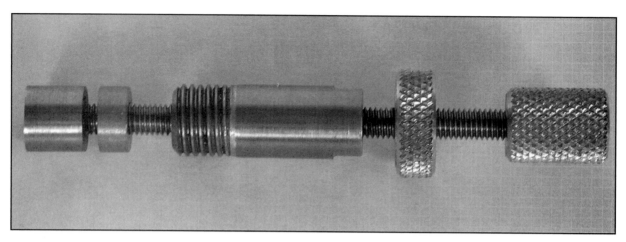

18. Assembled parts of adjusting plug.

19. Flats being machined.

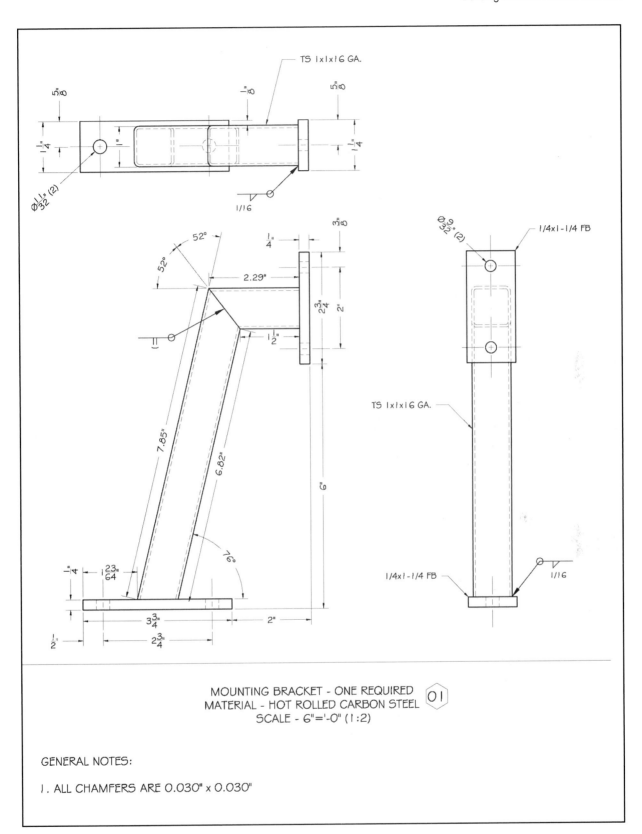

TS 1x1x16 GA.

Ø$\frac{11}{32}$" (2)

1/16

52°
52°
2.29"
7.85"
6.82"
76°

Ø$\frac{9}{32}$" (2)

1/4x1-1/4 FB

TS 1x1x16 GA.

1/4x1-1/4 FB

1/16

MOUNTING BRACKET - ONE REQUIRED
MATERIAL - HOT ROLLED CARBON STEEL
SCALE - 6"='-0" (1:2)

GENERAL NOTES:

1. ALL CHAMFERS ARE 0.030" x 0.030"

Drawing No. 6.

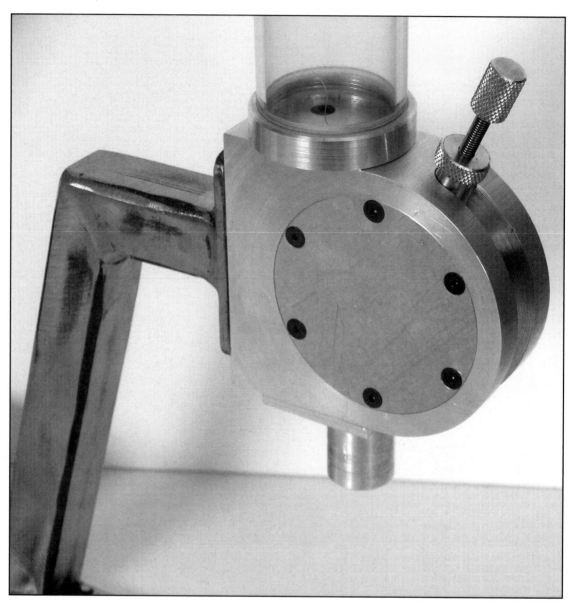

20. Mounting bracket.

powder as well as electrolytic reactions between brass and aluminum. If you have access to a source for anodizing aluminum this would be a good place to use it. Any metal material, and many synthetics as well, will probably work as alternates for almost any part of this project but I suggest that, regardless of material choices, it be disassembled and cleaned from time to time and propellants never lbe eft in it while in storage.

This was a fun project. And I think it is superior to other powder dispensers I have used in a couple of ways. It can be made to work for the left-handed operator simply by switching the bottom cap and the dispensing nozzle and turning the body over. And the knurled operating knob is easier for me to use than the lever I have found on other powder measures. And since it cost about ten times as much as a Lee or an RCBS then it must be ten times as good. Right?

Chapter 9

Some Thoughts on Housekeeping and Space Management

Sorry, NASA. The space management being discussed in these pages is not intended for you specifically although there may be some material here which you would find useful. The space management I am referring to here is the space we have available in our shops. It is a given that we, none of us, have enough space in our shops. My theory is that the space required for the operation of a machine shop, whether it be a hobby retreat or a 150,000 square foot manufacturing facility, is approximately 10% more that what is available. It doesn't matter how much room you have, you need 10% more in order to be comfortable.

And as for housekeeping — simple. Were it not for the fact that the talented and capable lady who is editing this work insists that two paragraphs do not make a chapter I could sum up housekeeping in one sentence and be done with it. You keep your shop clean by never letting it get dirty. It is as simple as that. I have little patience for the excuse that "I don't have time to clean up because I have too much work to do," and I have no patience at all for those who say that, "You can't keep a shop clean and get any work done." That is an excuse for laziness as far as I am concerned. But there are a few things you can do to make keeping your shop in an orderly condition a little easier. One of the things is the method in which you install your machine tools.

1. A good method for making leveling screws.

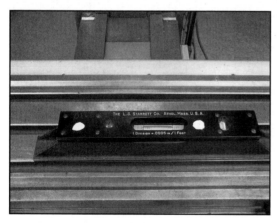

2. Leveling the milling machine in the X axis.

Housekeeping may not be the primary consideration when you install your machine tools but this method adds to the appearance of a neat shop and that never hurts. I learned about it in an oil tool manufacturing facility in California where I worked for a while in the early 1970s and I have used it to my advantage ever since then. Here is how you do it.

The example I am using is a Lagun FT-1 milling machine which I am installing here for what is probably the eighth or ninth time. My wife says I move my machines around more than she does her furniture but I don't know about that. Photo 1 shows one of the leveling screws fabricated from a 1/2"-13 SHCS, a 1/2"-13 nut and two pieces of 1/2" X 2" flat bar. Drill a clearance hole through the center of one of the flat bars and weld the nut to it over the hole. An alternative is to drill and tap 1/2"-13 through the material. Place one of these at each mounting hole of the machine. Another useful trick which I will sometimes employ is to put three or four layers of asphalt shingles or plain tar paper underneath each leveling pad. The tar paper conforms to the minor irregularities of the floor or to the bottom of the machine and

3. Left, putting down the grout around the base.
4. Right, grouting finished.

will help to keep the machine level and stationary over the long term.

Then level the machine. Photo 2 shows the machine level, or nearly so, in the X axis. Level the machine one axis at a time, making sure that the screws are all in firm contact with the bottom pad. After one axis is level, go to the other and make that level. Then return to the first and re-level. Do this as many times as is necessary to make the machine perfectly level in both axes.

Photos 3 and 4 are of the grouting process which is the key to this method. Mix the concrete, one of the "just add water" ready mix bags from the hardware store is what I use and one 40 pound bag will be more than enough for most small machines. Push the grout under the machine all around and build up a small flat or convex mound of the material. After allowing a few days for the concrete to cure paint it as shown in Photo 5. Re-check the level periodically as you work.

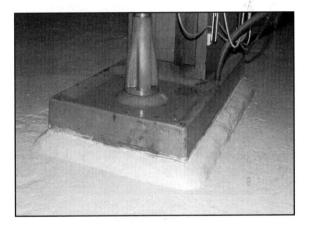

5. The finished installation painted. This makes it neat and easy to sweep the floor.

I know what you are thinking here. Why doesn't he bolt the machine to the floor? Good thought. And some machines must be bolted down. I have a 25-pound Little Giant® power hammer that has to be bolted to the floor or else I would have to have a seat on rollers so as to be able to follow it around. A horizontal shaper is another machine whose motion can cause it to move around. But most small machine tools, if carefully leveled on a stable floor, will stay there. If you have a punch press sitting right next to a surface grinder then they should both be bolted to the floor although you should probably move one of them to the other side of the shop.

I have nine machines in my shop leveled using this process and the last time I checked they were all still as level as when I set them up. And now when I move them all I have to do is break up the concrete and move them and then repaint the area where they were. Nothing to it! Plus, when I am changing inserts on a turning tool, and I drop the shim or the screw, it will not hit the toe of my shoe and go under the headstock of the lathe.

Things are much better now with the new synthetic coolants. But I have worked in shops where putting a dead steer in the sump would improve the smell.

6. Kurt vise, index head and rotary table mounted on one wheel carts underneath the work bench. Note the swivel base for the vise bolted upside down over the rotary table.

Another important, and often overlooked, facet of space management in a shop, whether it be a hobby shop, a neighborhood repair shop or a maintenance facility at the biggest industry in your part of your state is that your space is cubic! A 30 X 50 shop has 1,500 square feet of floor space. But if you have the usual minimum floor to ceiling space of 8' your actual available space is 12 thousand cubic feet. Granted you cannot use every one of them but with a little imagination and a few visits to the secondary material storage facility, known by some of the unenlightened as a scrap bin, you can transform some of that cubic space into valuable storage room. Photos 6 and 7 are pictures of my famous one-wheeled carts. By using a couple of swivel casters and some simple hinges I made places to store an indexing head with its footstock, a 6" Kurt swivel vise and an 8" rotary table, all in otherwise unusable space underneath a work table. Another indexing head lives under another table in Photo 8. All of these attachments are located in the area which is covered by my small bridge crane and can be retrieved, used and returned to their storage place without me having to lift more than the 12" Crescent wrench used to bolt them to the milling machine table.

Material storage is a tough one. Standard material lengths are generally 10' or 12' lengths for CRS shapes, 20' for structural steel and for HRS material, 21' for black pipe and 4' X 8' for sheet and plate material, expanded metals, and other materials usually bought in sheet form. So you need to have space available for storing these sizes and shapes. Now if you know beyond question that every job you do will require a piece of material at least 8' long

7. Index head swivelled out for lifting.

then you can sell as scrap every piece you have that is shorter than 8'. But in my shop, and I will bet in yours, that isn't the case. I try to be prepared to respond to every job brought to me, within reason, and I consider the only scrap that I generate to be the chips and swarf from my machines. Everything else has to be saved and if you save it you have to have a place to put it. The photos here show some of the storage places I use. I have storage places under the welding table, against the back wall, in the corner in the back room, under the saw table and anywhere else I can find space; floor space, wall space or even sometimes ceiling space that is not being used at the time for something else.

8. Another index head stored under a bench.

Raw materials are not the only thing you must consider here. Tools, hardware and a place to keep that extra can of coffee you bought the last time you were at Sam's Wholesale or at CostCo also have to be stored and they all have to be stored in a place where you can remember where you put it. Even in as small a shop as mine there are countless opportunities to lose something that you know you still have. You just can't find it right now! And face it. There will come a time when you have to (sob) throw something away. My grandfather is quoted as having said, "You will need everything you have at least once in every seven years." I have some things in my shop which have been here for 47 years and I still can't throw it away. That may be a problem for whoever settles my estate but right now it isn't one for me.

And one other note about hardware. Fasteners, snap rings, keys, hose clamps, dowel pins and on and on and on and on. The best way of all to manage this kind of inventory is to be near a hardware store or an automotive supply house. Go get six 5/16" X 1-1/4" bolts when you need them and do the job. But it accumulates. I attend auctions where just about everything you can imagine is for sale at one time or another. I bid off once, with a friend, 1,500 pounds of surplus aircraft hardware and I still have probably 500 pounds of what was my share of the loot. All of this stuff needs to be either stored where you can retrieve it when you need it, or disposed of. And who in their right mind would throw away a perfectly good, odd-sized, bastard thread bolt?!? So make a place to keep it. Luckily, many of the auctions also turn out to be good places to buy storage bins and cabinets and that evens out the playing field. I suggest you get a good tape embossing machine and label the drawers. It is time well spent.

Other examples of using vertical space are shown in the photographs "A" through "O" which accompany this chapter. The welding electrode storage cabinet is the top part of an old refrigerator which is installed near the ceiling giving room for a tool grinder underneath. Old refrigerators are a little over rated as storage for electrodes because they will not keep low hydrogen welding rods at the storage temperature recommended. But with a 40 Watt light bulb they will keep the moisture

The sound of $500 worth of insert, boring bar, chuck jaws and the customer's job becoming, in an instant, $2.75 worth of scrap is one that will stick with you.

to a minimum and if you are not doing code welding that probably will suffice. I cut the freezer compartment from an old refrigerator and installed it high enough on the wall to make available for other purposes the space underneath it. TIG wire is kept in airtight containers made from PVC pipe. It is out of the way. It is convenient when it is needed. And it is protected from oxidation by being sealed up.

Every shop is different and the tools used in every shop are different but the approach to keeping them neat and orderly is the same. It is well worth the little amount of time it takes to prepare places to store tooling attachments, tools, materials or any of the thousands of items it takes to operate an efficient shop, whatever kind of shop it is.

Try to avoid if possible deep bins and drawers. If you keep odds and ends of materials in a deep box you will find that the only pieces available to you are the ones on top. I used to keep short pieces of aluminum leftovers in a barrel outside. I punched holes in the bottom so that rain water wouldn't accumulate in it and it held a lot of aluminum. But I may as well have had all but the top six inches of the barrel filled with sand because I never dug down beneath the first layer to look for a particular piece and dumping the whole thing out was just out of the question. The same thing is true of the bottom drawer of many storage bins where the odd fastener is tossed. There is a rule of diminishing returns to be followed in keeping unsorted piles of fasteners around. I am contradicting my own words here when I say that sometimes things should be disposed of, but it is a sad truth. Grit your teeth and get rid of it.

And back to the one-liner about housekeeping. Keeping your shop clean by never letting it get dirty. In the now famous book, *All I Ever Needed To Know I Learned In Kindergarten,* by Robert Fulghum, he says "If you make a mess, clean it up." And that is what it takes. Develop the habit of putting tools away at the end of the day. Even if you are going to use the 2"-3" micrometer first thing tomorrow morning don't leave it out on the work bench tonight. Put it away and get it out again in the morning. You will find that you spend less time looking for tools and your tools will probably last longer as well. Sweep up the chips at least once every day instead of just on your birthday. I try to keep my shop so that I never have to apologize to visitors for the way it looks. There are times when I am machining a big job from aluminum and it looks like an aluminum scrap car has crashed. But come back in the morning and the chips are in the chip barrel and the tools are back in place and the coffee is made and all is right with the world.

I do have one advantage in keeping my shop clean that some of you may not enjoy. Her name is Nan. I couldn't keep the shop as clean as I do without her help. I make the joke that my wife helps me to keep the shop clean this way. She says, "Sweep that up!" and I go and sweep that up. But the truth is that we work together keeping both my shop and her house clean and we both like it that way.

I need to find a place that will sell me just one glove. I use up left hand gloves welding at about four times the rate I do the right ones and I hate to throw away a good glove.

A. Above, a storage shelf for short pieces of alloys. Note the marking on the ends of the material. This will help to avoid the appearance of "mystery metal" in your shop.

B. Left, a couple of shelves of mostly structural steel under the welding table.

C. Below, these shelves are under the roll table which serves the cutoff saw. I can store pieces from about 12" to 18" in length here.

It can only be considered work if you had rather be doing something else.

D. Above, a shelf in the back room for short, heavy materials. Some aluminum and also some pieces of synthetic materials.

E. Above right, more of the shelf in the back room.

F. Right, rack of odds and ends made from some scrap PVC pipe.

G. Below, the long stuff. Notice the 4' to 8' lengths of shafting leaning against the wall on the right.

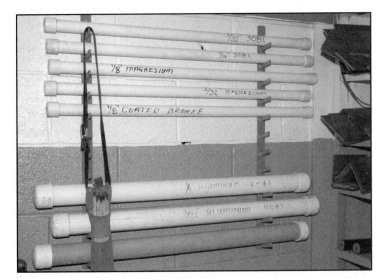

H. Above left, a good place for 36" lengths of welding and brazing wire.

I. Above, the power work head and an assortment of chucks for the tool and cutter grinder.

J. Left, back corner of the shop. All of this material is stored without using more than a couple of square feet of floor space.

K. Below left, grinding wheels and tool steel.

L. Below, another use for short pieces of PVC pipe. This is behind the surface grinder.

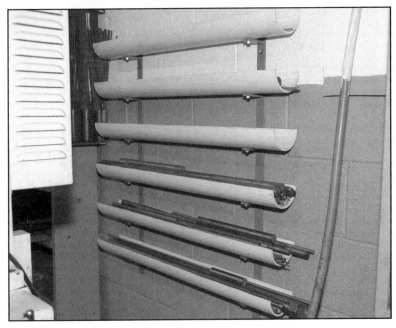

M. Top, shelves and bins in the back room. I had to drink a lot of coffee to create this storage facility.

N. Left, another view of the back room. You can see a corner of my next machine rebuilding project in the left foreground.

O. Right, some of the bins and cabinets I have purchased at auctions.

Chapter 10

Introducing the Best Tool Caddy
in the World

Perhaps I should qualify the title of this chapter by saying that this is, at least, the best tool caddy in MY world. Your world is probably a little different from mine and therefore your tool caddy will be a little different. But if your approach to your work is at all like mine you will find that a well thought out, conveniently arranged tool caddy will make your world a little better. My tool caddy includes the tools that I most frequently use in the operation of my lathes or milling machines. A tool caddy for a grinding center would have to include wheel dressers, magnetic parallels, spindle wrenches and so on. If your work requires that you use post hole diggers frequently then, by all means, include a place to keep a set of post hole diggers. The point is this. Decide what tools you would like to have always on hand without going to the tool storage cabinet against the wall over on the other side of the shop and incorporate them into your tool caddy.

Why is it, I wonder, that the designing engineer will happily share the credit with the machinist/fabricator when the part works but will never share the blame when it does not?

My tool caddy is a work in progress and will ever be. When I bought a 17" lathe a few years ago I had to get a bigger tool post. The tool blocks you see along the top of my caddy in Photo 1 were the ones I used on the 14" lathe and the ones below the white shelf are the ones I use on the 17" lathe. So the caddy changes as the machines it serves change. Some days I use the lathe and some days I am at the milling machine but nearly every day I will use the caddy. Except for the days when I have to work in the yard and on those days I have to content myself with thinking about my tools. The caddy has no provisions for lawn mowers and rakes.

1. Front side of the caddy.

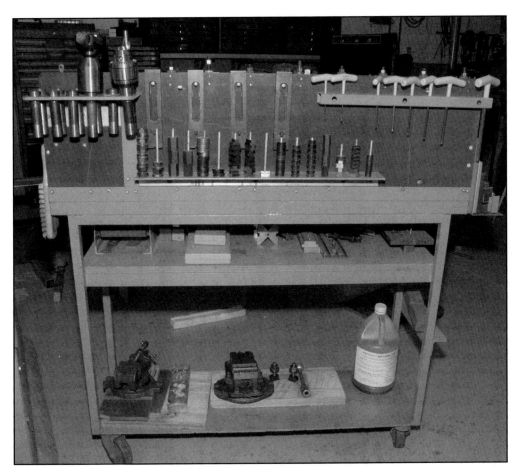

2. Back side of the caddy.

The photos accompanying this article show some of the things I have done but are only suggestions as to how you may want to do it and by the time you read this, I will probably have changed it anyway. As a matter of fact, if you will look closely at the photo at the beginning of this chapter and compare it to Photo 1 you will see changes. Photos 1 and 2 show the front and back of the caddy. The front shelf was made from a piece of Formica® covered kitchen cabinet top saved from a remodeling project in the kitchen a few years ago. The rest of the caddy was made from odds and ends of materials left over from other jobs. (Notice I did not say material from the scrap pile. I do not have a scrap pile because until material is reduced to chips or slag it isn't scrap.) The casters had been on a shelf in the back of the shop for years, waiting for just this occasion. The protective covers on the end mills and cutting tools are not there to protect the cutting edges of the tools. They are to protect ME from those cutting edges. On the back side and on the left end is located a variety of hardware for attaching jobs to the milling machine table. Also found on the back are storage racks. One is for R-8 collets and the other for hex wrenches.

The most recent addition was the holder for the Jacobs chucks and the quick-change tooling shown on the left side of the caddy. They were mounted on a bracket which at one time held the display for a DRO on the milling machine but are more conveniently at hand where they are now. I can remove the entire unit and mount it on the milling machine table as

If you are a horse you probably don't need to wear hard toed shoes in a machine shop. Otherwise, it is a good idea.

3. Quick-change tooling moved to the milling machine table.

4. Quick-change tooling back in its place on the caddy.

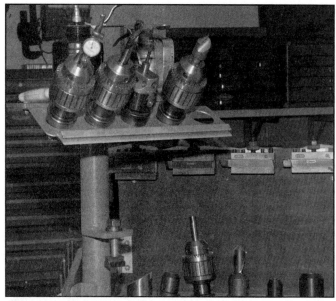

shown in Photo 3 when a job warrants it. The job which I was working on at the time I took these photos is a good example of how this attachment is useful. I use a quick-change spindle attachment on my milling machine most of the time and on the job you see in the photos, I was required first to locate the part so that center of the large hole was at the origin. To do this I used the Last Word® dial indicator mounted in the first chuck on the left. I then drilled two holes with a 31/64" drill, bored them out to .503/.505" with the boring head, hit them with a counterbore to break the corners, and machined a 3/8" X 3/8" slot across the middle of the part. These parts are locating fixtures for a customer of mine and there was a series of them to be done. Having the required tools right there within easy reach saved a significant amount of time over the course of the job. I did not have to stop the spindle to change the tools with the exception of the dial indicator. When the job is done and the tools are no longer needed at the milling machine they go back to their home on the caddy as may be seen in Photo 4.

The calculator in Photos 5 and 6 was a favorite pocket calculator I used for years until it almost literally came apart in my pocket. I made the

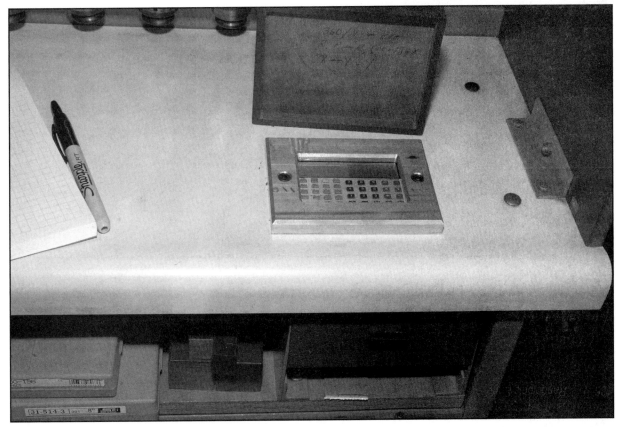

5. Pocket calculator uncovered.

aluminum plate to hold it together and the phenolic cover plate and now it is always at hand. The formula written in the top of the cover is one for converting polar co-ordinates to rectangular co-ordinates when machining bolt patterns. But I now have a DRO which takes care of that sort of detail.

On the middle shelf are a pair of V-blocks, a knurling tool, soft-jaws for the 3" Kurt® vise, some odds and ends of setup tooling and parallels and, on the right hand side and shown in Photo 7, is a drawer with a smaller set of V-blocks, a Starrett® screw jack and some other odds and ends of setup accessories. I rescued the drawer from a Kennedy® tool box which lost a battle with a fork lift in a shop I worked in at one time. The *Machinery's Handbook* drawer was the only part of the tool box which survived.

Items on the bottom shelf are a gallon container of cutting oil, a couple of milling machine vises and some useful pieces of wood.

When you build your tool caddy some things you will want to consider are:

1. How big should it be? You need to be able to move it conveniently from one machine to another in your shop so it will have to be of a size to go between the band saw and the welding table. It also needs to be big enough to be useful. If your shop is so small that you only have room for a one foot square tool caddy then you probably don't need a tool caddy.

2. Which machines will it serve? Only you can answer this question. The machines which most commonly will benefit from this

No matter what the product at the factory, from jockey shorts to freight cars, somewhere, either there or nearby, you will find a lathe and a milling machine and somebody who knows how to use them.

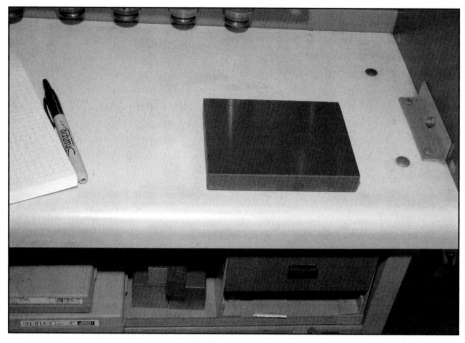

6. Pocket calculator with cover in place.

7. Drawer for setup fixtures.

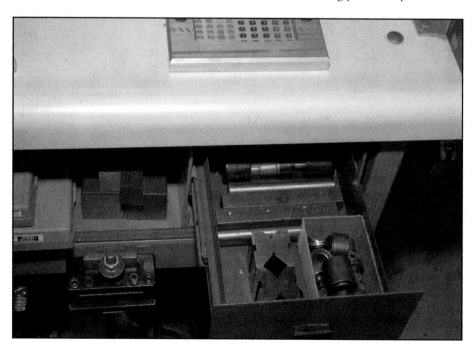

approach will be lathes and milling machines, but a tool caddy with TIG nozzles and tungstens or MIG contact tips plus a place to keep a variety of welding clamps and torch tips may serve well if your work includes a lot of fabrication or welding repairs.

The key to a useful and convenient tool caddy is to know where to stop. A good tool caddy does not replace your roll-around tool box. You bring the tools and measuring instruments for a particular job from the tool box to the caddy and when you are finished they go back to the tool box. If you are turning diameters between .450" and 1.360" then you will need the 0-1" and the 1-2" micrometers and the larger mikes will stay in their nests in the middle drawer of your roll around. If you are cutting threads get out the thread gage and the center gage while the feeler gages and the set of radius gages stay at home.

Not everyone will benefit from nor even want a tool caddy. Many people who build them, including yours truly, build them because they love making things that make their shop more appealing to themselves if not to others. It is one of the prices we pay for having a place where we are truly happy doing what we do just for the sake of doing it. And my tool caddy will never accommodate any lawn tools of any kind!

Chapter 11

Another Wife Pleaser

I know I am flirting with disaster here by referring to this project as a "wife pleaser." But I have already admitted to being old and set in my ways when it comes to recognizing that the ladies have as much a place in the shop as do the men. This project could be called a "husband pleaser" or a "Granny pleaser" depending upon the person you have in mind as you are making it, but it is a sure thing that somebody will be happy with it if you take the time to make a similar piece of furniture in your shop, for your house.

Young people these days are sure adventurous. I am amazed at the number of young couples who will get married and then try to set up housekeeping without even owning a welding machine or a lathe!

Whoever builds it for whomever, this is a nice project which doesn't take much time and the concept can be used over and over in many places throughout your home or even in your shop. Photo 1 is a picture of a tool chest which my father bought in 1940 from a local cabinet maker. Another machinist in town had hired the cabinet maker to make it for him and for some reason didn't take it when it was finished. My father paid $5 for the chest and used it until his retirement in 1969. I have my own tool chests fitted with my own tools and really did not need the chest in my work and, of course, selling it or disposing of it in any other way was out of the question. So I made an end table out of it to use in the living room. The perfect solution!

A similar story can be told about the bedside table in Photo 2. A place where I worked at one time purchased a set of pin gages in diameters from .011" through 1.000" in .001" increments. There were 990 individual gages in the set. They were good gages and I will bet that they have not been used more than a dozen times in the past 20 years, but that is beside the point. They were delivered to the shop in two nicely fitted cabinets, but shop management at that time, or rather shop mis-management as the case may be, had purchased some Stanley-Vidmar high density storage cabinets and the gages were removed from their original cases and placed in the new cabinets. When the original cabinets were thrown away I retrieved them and made two of the bedside tables as shown here.

The third example, see Photo 3, started life as a tool chest owned by my grandfather and was one of the few

1. End table made from tool chest.

things to survive the fire when his shop burned in 1935. By making one of these ornamental iron table stands it was turned into a great chest for silverware and serving utensils. I really came out looking pretty good to the kitchen staff when I finished this and took it in for approval.

So with all of these examples as precedents I don't see how you can possibly resist making one of these bases for yourself or for your wife, or even, if it applies, for your husband.

The project in this example will be to make a base for a crate which my son found and bought for his mother. It was once filled, but is sadly empty now, with Shiner 96 ale. Whatever you will use as a top for this job is immaterial. The base is what constitutes our project here and you will have to provide the reason for making the base. Which will influence directly the dimensions of the base. I am not providing a materials list for this chapter because the variables are so great in number. The things you might be interested in here are the methods for joining the corners, twisting the square steel, and making and attaching the feet. These things can apply to almost any size table or table base.

2. Gage cabinet made into bedside table.

We said back in Chapter Seven when we were working on the fireplace screen that we would discuss the twisting of hot rolled steel in further detail and now is the time for that discussion. It doesn't take as much force to twist a piece of HRS as many people think. I have twisted hundreds of poker handles using the method shown in Photo 4.

3. Ornamental iron stand gives grandfather's toolchest new life.

Clamp it in a vise and get out the big adjustable wrenches. I will still do this today if I only have one or maybe two pieces to twist or if my twister has been rolled back into the corner and is hard to get at. But if I have many pieces to twist, or if I have to twist parts bigger than 1/2"-square, I use one of the other methods shown here. I once made a decorative fence for a customer and there were over 200 pickets made of 5/8" hot rolled square and each one of them had two twisted sections in them.

I made the fixture you see here in Photo 5 to attach to the compound of my lathe and with this fixture, a four-jaw chuck, and a 7-1/2 HP

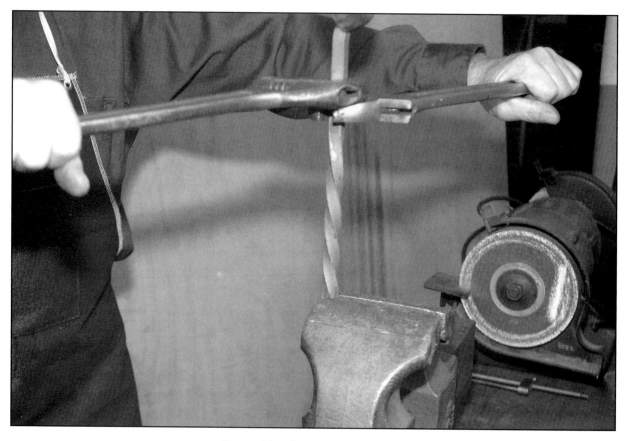

4. Method of twisting hot-rolled steel.

spindle capable of turning at a speed of 35 RPM, I was able to do the job in about a day's time. Twisting these by hand would have been out of the question and twisting them hot would have taken far too much time. I try not to take on jobs of this size anymore, but if I ever do I still have the fixture and the lathe. One word of caution here. I do not recommend using your Aloris tool post for a twisting fixture. It will stand one or two applications or maybe a few dozen but if you have a lot of parts to twist and plan to do it on your lathe then make the fixture.

5. Fixture made to attach to compound of lathe.

Heating the steel in a forge or with a torch can work well but there are some drawbacks.

Photo 6 shows two examples of steel which have been twisted hot. The 1/2" square shovel handle and the 1" square center post for a tool stand were both heated in the forge and then placed in a vise and twisted with a wrench. You can see the uneven twist which was caused by the steel being cooled at either end from coming in contact with the vise and with the wrench. I will have to say this. When it comes to twisting 1" square steel in my shop doing it cold is not an option. Uneven twists or not!

The best way of all is seen in the photos (Series A) of the twisting machine I designed and built back in the 1970s.

This machine has twisted over 8,000 parts counting pokers, shovel and broom handles, railing pickets and fire screen handles. It was constructed of odds and ends lying around the shop including an old gear reduction unit from a coal stoker and a 1-HP motor. It is made so that the operating lever engages the drive when moved in one direction but will act as a positive stop when moved in the opposite direction. This is necessary when you want to stop the twist at an exact point of rotation.

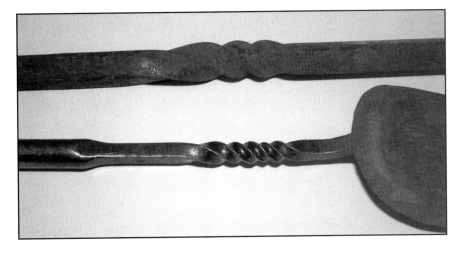

6. Two examples of steel that has been twisted hot.

Whichever method of twisting you use, assuming that you choose to twist the cross pieces at all, be sure to make the twists so that they do not look out of place or out of proportion. In this example, I have chosen to put a 6" twist in the long parts and a 3" twist in the short parts. This requires that I make the 6" twist two full turns or 720 degrees while the shorter parts have only one full turn, or 360 degrees. The result is that the two twisted members have the same appearance. If you use other fractions then apply the arithmetic and make the twists to match. A 4-1/2" twisted area would have one and one-half turns and so forth. This attention to detail will make the finished product much more appealing to the eye.

One other point about twisting hot rolled steel for decorative purposes. If you are an absolute perfectionist, you will want to know that a piece of 1/2" square steel twisted two full turns will be shorter by .025." This will make no appreciable difference in most applications, but you might have a job where it will matter, so keep that in mind.

The mortise and tenon method used here for fastening the steel cross pieces to the vertical legs is really worth the time that it takes to do it. The photos (Series B) show how this is done. This example is for fitting a 1/2" square cross member to a 3/4" square leg. Machine the cross member to a diameter of 3/8" of an inch. Then drill a 3/8" hole in the vertical leg at the required point. Countersink the vertical member to provide a good weld and weld only on the end of the cross member.

Grind the weld smooth and you have a secure joint with no visible welds. There are other advantages to this method. You can drill the holes at the exact location for the cross member which makes accurate assembly easier. You will find that using this method will result in a faster assembly because you don't have to clamp and measure for each weld.

The micrometers which will measure in both inch and Metric systems represent a good try. But I still contend that trying to use both measuring systems at the same time will get you into trouble.

Series A: Doing a project with a twisting machine.

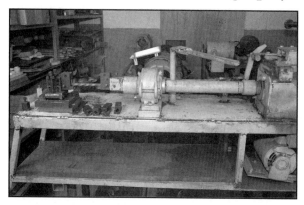

A1. Twisting machine with dies for twisting 3/8" square, 1/2" square, and fire screen handles.

A2. Two full twists, or 720 degrees, in 6". Material is 1/2" hot rolled square steel.

A3. Before twisting. 3" long to receive 360 degrees of twist.

A4. Fire screen handle before twisting.

A5. Fire screen handle after 1 full twist.

Series B. Mortise and tenon method for fastening steel cross pieces to legs.

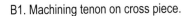

B1. Machining tenon on cross piece.

B2. Mortise and tenon before assembly.

B3. Mortise and tenon assembled.

B4. Top frame clamped and ready to weld. Notice beveled weld preps.

B5. Assembly clamped and ready to weld.

B6. Above, grinding facets on the feet.
B7. Right, above, round mortise and tenon parts before assembly.
B8. Right, bottom, 1/2" round and 3/4"round parts assembled.

7. Completed project.

I prefer a casual compliment, or criticism, about my work from someone whose experience and abilities I know and respect, to raves and accolades from someone who has no knowledge of the work.

The time you save by not having to clean up the welds and filing and grinding the joints will make the method worthwhile.

As you can see in the photos this method will work as well with round stock as it does with square. The chief difference between joining square parts and round parts is that you have to drill and tap into the end of the cross piece to install a tenon. The mating surface will require machining with an end mill of the appropriate diameter.

You can make both vertical and horizontal pieces of square stock, or of round stock, or one of each. Square legs and round cross pieces or the other way around. It takes a little more time but it is neat and accurate and you will get a little of that time back when you grind your welds. The photographs included here should make the process clear, but, as usual, there are some things to remember. Make your tenon diameters proportionate to your material cross sections. In this example, the vertical pieces or the legs are of 3/4" square stock and the cross pieces are of 1/2" square stock.

If you make the tenons 3/8" diameter you will have adequate strength in the tenon and you will still have enough material left after drilling a 3/8"-hole through the leg. Countersink the side to be welded so that you do not run the risk of grinding all the weld away.

The feet for this little stand are simply beveled pieces of steel, sawed either from 1/2" X 1-1/2" flat bar or from 1-1/2" square stock. Grind the bevels as shown in the photo and drill a 1/2" hole through the center so as to be able to attach them in the same way the joints are made.

Photo 7 is a picture of the completed project. This job didn't take long, maybe four to eight hours, including the painting, but it will depend upon how much tooling you have to improvise. It is a good example of a job that brings pleasure to the craftsman doing it as well as to the other folks at your house who don't spend their time in the shop. Wife or husband!

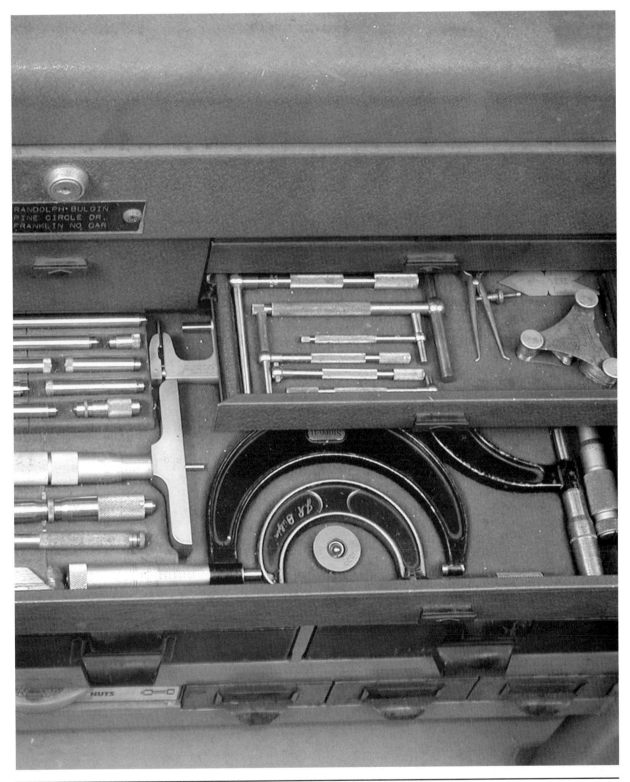

Chapter 12

Organizing Toolboxes

This chapter is another of my efforts which was accepted and published by *Home Shop Machinist* magazine. It first appeared in the May/June issue of that magazine in 2005.

"A place for everything and everything in its place." We have all heard that and most of the time it works. I grant that it is harder to do with some things than with others, but, in one form or another it is generally a good rule. The method of preparing a "place for everything" shown here works better for some tools than for others. There may be some other obsessive-compulsives besides me out there who will be tempted to make storage compartments of this type for their shovels and hedge trimmers, but differences in practice are like differences of opinion: They make the world go around. My world seems to go around more efficiently if I have all, or most, of my precision measuring tools and gages stored in such a manner as to protect them from the damage which can result from piling them into a drawer, one atop the other. An added bonus is that it frequently saves me the time I might otherwise spend looking for these tools. That is, if I remember to put them back in their nests after I have used them.

I once worked with a machinist in a large shop who treated his measuring tools in the same way he treated his wrenches and hammers. They were piled into the drawers of his tool chest one upon the other and when he closed his box at the end of the shift tools were thrown back into whichever drawer was open even if it wasn't the same drawer they were found in at the beginning of the shift. He was, or could have been, a capable and knowledgeable machinist. But his work suffered from the condition of his tools and everybody in the shop knew it.

Buy for yourself a GOOD boring bar. Then hide it. Don't let your wife know where it is. Don't let your friends know where it is. Don't even get it out and use it yourself except for the job which requires a GOOD boring bar. You won't regret it.

1. Place tools in order you want.

So without discussing further the philosophy of tool care, let's do something about it. The methods and materials shown here are but one way of doing this. Styrofoam®, foam rubber, corrugated paper and even wood can be made to work as a bed for holding the tools but the best material I have found is Fome-Cor®. It is a product used in framing pictures and two nice ladies of my acquaintance who operate a frame shop gave me the scraps to do the project shown in this description and these photos. Fome-Cor® is

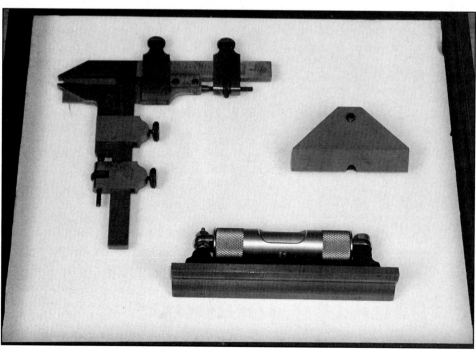

available in various thicknesses and what I used here is .200" thick and is easily cut with a sharp knife.

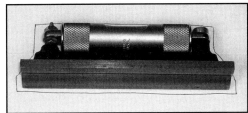

2. Draw outline around tool that will occupy full depth of drawer.

Begin by cutting as many pieces as you need according to the depth of the drawer you are working with. The shallow tool drawers of a machinist's tool chest will require three thicknesses of the .200" Fome-Cor®. Cut the pieces so as to leave a space around the edges for tucking in the felt. Then place the tools on one of the pieces in the order you want them to be stored as shown here in Photo 1. Leave enough room between the tools to allow for the thickness of the felt or other material you may have chosen for the final lining. I have chosen here for demonstration purposes three tools of varying thickness. Draw an outline as shown in Photo 2 around any tool which will take up the full depth of the drawer, in this case the 4" machinists' level. Keep it as simple as will accommodate the full outline of the tool and make it about 1/8" larger than the tool on all sides. Cut out the outline of all tools which will fit this first, or bottom, layer.

3. All pieces cut out.

Now place the piece you have just cut on the second layer and repeat the process. Trace around the inside of the first profile you cut and draw an outline around whatever tools will only penetrate to this layer. In this case the home made dial indicator attachment for measuring hole depths. Again cut the outlines slightly larger than the tool and make the outline around the machinists' level about 1/8" larger on all sides than was cut on the bottom layer.

Continue this process until you have cut all of the pieces to fit. Cut a bevel around the inside of the outlines on the top layer to help keep the felt lining more free of wrinkles. Photo 3 shows all of the pieces cut and Photo 4 shows them stacked in the drawer ready for the lining. I use felt as a lining although I have seen Naugahyde® or upholster's leather used to good effect.

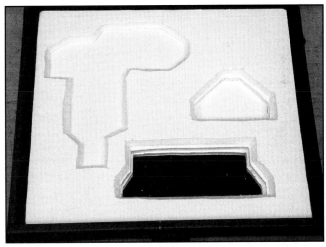

4. Pieces stacked, ready for lining.

Glue the three pieces of filler together and allow them to dry. Then place a piece of the lining material over the outlines. Remember that the lining required will be larger than the drawer size. Depending upon the depth of the drawer and the number and size of the tools you are working with it can be as much as 3 or 4 inches larger so don't scrimp when you select your material. Place each of the tools in their respective locations and press them firmly down into their beds. Start with the thickest, in this case the level, and seat each tool before beginning with the next. See Photo 5.

Trim the lining around the edges, leaving from 1/2" to 5/8" on all sides. Cut a notch at each of the corners to help avoid wrinkling, then place the entire assembly, tools included, into the drawer and

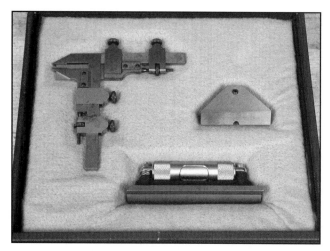

5. Seat each tool before going to the next one.

tuck the edges down all around using a table knife or a machinist's scale. Add this to the list of thousands of uses for the scale!

Photo 6 is of a couple of drawers in my everyday toolbox that I made while working in an oil tool manufacturing facility in 1972. It shows some of the effects of the passage of the intervening 33 years, but it still serves the purpose for which it was made and if I can keep the tool box out of the way of fork trucks and train wrecks for a few more years it will have served me well.

Don't try to put too many tools into one drawer. Forming the lining gets more difficult as the density increases and most tool chests have enough room in them to accommodate all the tools you will want to go to this much trouble for. If you run out of space for the proper storage of tools in one machinist's chest you will have the perfect excuse to justify the purchase of a second. Or a third or fourth, even!

I have had visitors to my shop say to me that I obviously spend all of my time cleaning and organizing. Not true. I spend time keeping my shop clean and I consider that to be a priority as will be discussed in Chapter 14, but a little time spent in organization makes it easier. My best and most profitable customer is a customer I got because he was impressed with the cleanliness and orderliness of my shop. I like to think I keep him as a customer because of the quality of my work, but in order to keep a good customer you first have to get a good customer and taking care of your tools helps with that.

6. Randolph's everyday toolbox.

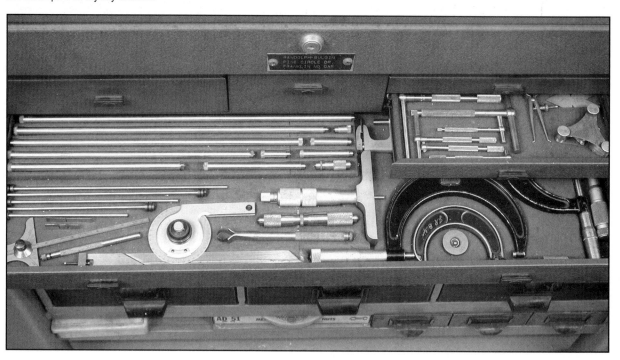

Chapter 13

Just Another Job

I support the unification of measuring systems all over the world. And the metric system is likely the most widely used globally. But I am too old now to make such a basic cultural change. My default system is the English system. I wake up in an inch world every day and then I convert. Thank Heaven for pocket calculators.

1. Worn track guide area.

I t isn't fair to call this "just another job" — for the simple reason that no job that is worth doing should be called "just another job." Sometimes we get into ruts and begin to think of jobs in those terms and that is when the work can start to become less satisfying. This was actually a pretty interesting job and it was done for a customer who knew the difference between repairing a part to just make it serviceable or really taking the time to do the job right.

And take my word for it, that type of customer is a good one to have. When rebuilding or reconditioning machinery, and this is true for almost any kind of machinery, whether it be automotive, woodworking, metal working or manure spreaders, there are some choices to be made. You can repair it and make it serve until a new part or a new machine can be obtained. You can repair it so that it works well but still looks repaired. Or you can take the time to repair it and make it look as if it had never been broken in the first place. When you get a customer who wants the latter option, do him a good job because he is usually a good customer. Sometimes the difference in the cost of a "just get it back to where I can use it" fix and a "fix it right whatever it takes" can be significant. And sometimes it isn't worth it! But knowing when to use which method can help you out in the customer relations department. The job we will discuss here in this chapter is such a job and, although you may never see exactly this job come into your shop there may be some information here which can be useful to you on other jobs. Now if it really is a manure spreader, you will have to decide for yourself how much of your resources should be spent making the repairs.

The DeWalt power tool company, which we now know mostly as manufacturers of portable construction tools — and pretty good ones, by the way — at one time was best known for the radial arm saws they made.

In fact, 30 or 40 years ago the term radial arm saw and DeWalt were synonymous. A radial arm saw was just called a "DeWalt" and people knew you were speaking of a radial arm saw. The overarm of the saw used as an example in this repair was probably made during the 1950s or 1960s. I am just guessing at that, but I won't be far off. The machines of that era were pretty sturdy. Lots of cast iron and good heavy construction. But hours and hours of use and, in this case, probably being backed over with a fork truck or being visited by some other catastrophe on the job, has taken its toll, and eventually the machine has lost its accuracy and its usefulness. The problems with this particular job were compounded when someone at one time or another tried to do one of the "just get it back where I can use it" fixes and left the part in pretty much of a mess. What I call "fixed at" rather than actually repaired.

2. Broken casting

3. Broken casting. Notice failed weld.

Photo 1 shows the damage done by many hours of use while Photos 2 and 3 show the most severe problem, the broken casting. You can clearly see why the term "fixed at" applies here. The goal here is to re-machine the spherical grooves where the saw head rides and to repair the broken casting. It makes no difference which area is addressed first.

4. Fixture made from angle iron

Know your material. If you smell cotton burning while you are welding A36 steel plate, your shirt is probably on fire.

5. Indicating elevation.

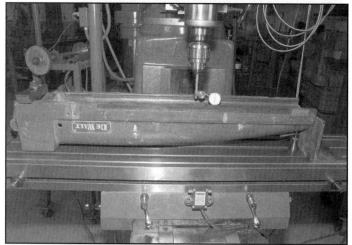

6. Showing side to side placement.

I began the job by setting the part up on the milling machine to re-machine the spherical grooves which support and guide the saw head. Photo 4 shows the fixture I made from a piece of structural angle iron. The other end of the part was held in place by bolting it flat down to the milling machine table. Photos 5 and 6 give examples of the process used for indicating the part true in the milling machine. The area which was to be cut was just over 25" long. It was possible, by using the small jack shown in Photo 7 to control elevation and by tapping the part from side to side with a lead hammer, to indicate it true to within .001" over its entire length. This equates to less than .0005" per foot which is pretty good for a machine of this type.

You do not want to take any more material off here than is necessary to restore the groove. Take light cuts, .002" to .003" per pass, and cut both sides in this one setup. I wound up taking about .008" from each side. The cutter used was purchased for this job and is a convex cutter with a 1" radius. Make sure you have finished both sides of the groove before you remove the part from the milling machine. Photos 8 and 9 show the cut in progress.

Now to the hard part of the job. Begin this portion of the project by machining a cylinder with an inside diameter equal to that of the original inside diameter of the overarm. If the damage

is too severe to get an accurate measurement then measure the column of the saw and use that dimension plus about .005" to .008." You are only going to need a small segment of this cylinder but you will have to machine the full diameter. If you do as I do you will save the remainder of this part for about 12 years and then throw it away two weeks before you get another saw to work on. All

7. Above, screw jack to control elevation.
8. Left, 1"-convex radius cutter.
9. Bottom, 1"-convex radius cutter on the first cut.

The true craftsman is not the one who never makes a mistake. It is the one who can make a mistake and then fix it before anyone knows about it.

10. Top, starting the cutout for the boss with an end mill.
11. Bottom, finishing the cutout with a boring bar.

There are some customers who absolutely cannot be pleased. The trick is to recognize them before they become customers.

a part of running a small shop.

Set the cylinder up in the milling machine and machine the cutout for the boss where the clamping bolt will go. Photos10 and 11 show the process. Start the profile with an end mill and then bore to the appropriate diameter with a boring bar. Be sure to cut deep enough to be past the centerline of the part.

The three parts you will need are shown separately in Photo12. Obtain the required dimensions for the boss by measuring the good side. Saw the flange from material of the appropriate thickness. If the flange you are replacing is 3/8" thick then saw the replacement part from 3/8" material. Different thicknesses would work as well but would be noticeable in the finished job. And don't forget, part of the goal here is to make the repair in such a manner as to make the part look like it was never broken!

Weld up the steel parts and fit them into place as shown in Photos 13 and 14. Grind the welding prep at this point on both the cast iron and the steel replacement part. Notice the studs inserted in the weld area in Photo 15. This is an old trick which will help you to weld cast iron parts whether you are welding cast iron to cast iron or cast iron to steel. If the latter is the case, as it is here, put the studs in the cast iron part. Drill into the cast iron and embed threaded studs into the part in several locations. Then weld around them and to them as you weld the joint. Welding cast iron is pretty tricky at best and this method will serve to make a stronger joint. Another thing to take notice of in this photo is the washers sandwiched between the two parts being welded. They are placed there to keep the parts in place in relation to each other while the 5/8" bolt clamps them together. Both pieces of the weldment are arranged around a piece of tubing the same diameter as the column of the saw and clamped firmly into place with the vise grip C-clamps. Spend all the time you need to insure that the parts are properly clamped into place relative to each other.

It's time to start welding. Warm the assembly up with an oxy-acetylene torch to about 350 degrees F before welding. There are several good cast iron electrodes available now. What is used here is a Eutectic product which is an electrode so expensive that you will save all of your rod stubs but it does a good job on cast iron. Whichever electrode you use it is a good idea to weld small segments and peen the welded area frequently. The chief cause of cast iron weld failure is the contraction of the deposited material causing the cast iron to crack adjacent to the weld. By peening the welded area while it is hot you will offset the effects of this contraction and will help to insure a good, long-lasting weld.

12. Three pieces required to make the missing part.

Finish the job by grinding the contour of the part back as nearly to the original profile as you can. The finished job is shown here in Photo 16.

13. New part tacked together.

You may never see a DeWalt overarm in you shop. But there may be some procedures covered in this chapter which will apply to the parts you may be brought by the fellow who is restoring an old tractor or rebuilding an old pump or repairing some other piece of equipment and he has heard around the neighborhood that you can fix things that nobody else can. That can't hurt, can it?

14. Above, new part ready to be welded into place.
15. Right, set up and ready to weld. Note studs in weld area.
16. Below, finished job.

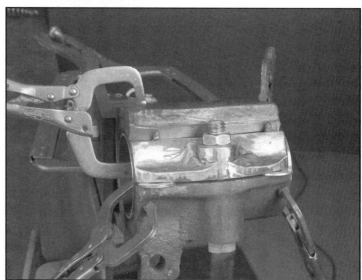

I don't understand the modern trend of body piercing. I have been trying to avoid that ever since I emptied my first chip pan. I probably have endured more body piercings than anyone you know, but none of them were done on purpose.

Chapter 14
Building and Erecting a Flag Pole

In the days following the tragic events of September 11, 2001, I, like many others in this country, felt moved to do something. I couldn't provide any help to the emergency workers nor to the families of the victims, but I felt like I had to do something. So I, again like many others, decided to display the American flag where I live and work. I knew it wasn't much, but at least I was doing "something". It is a little embarrassing when I realize that it took such a malicious attack on our country to prompt me to do this, but better late than never.

Sadly, many of the folks who were likewise moved to display the flag rushed to put flags up and there they hang to this day, neglected, ragged and forgotten. So I make this request. If you choose to follow the plans in these pages for building and erecting a flagpole, then take the time to learn how and when to properly display the flag and do it right. Otherwise, your time will be better spent building a fence. Go to the website located at www.usflag.org/us.code36.html and you will learn the things you need to know in order to display the flag properly. And you will also have the satisfaction of building a classy pole to fly it from.

The bill of materials attached to this article is only binding if you build a flagpole exactly like the one I built. Which is a pretty good flagpole, by the way. But here comes the inevitable disclaimer. The soil where I live, in the mountains of Southern Appalachia, is a solid, stable clay soil. If you live

It really irritates me when I run into people who would have you to think they know everything. It makes it so much harder on those of us who really do.

1. Leveling the base plate.

Materials list: Flag Pole

1. Whatever goes into the ground. See the disclaimer in the text of the article.

2. (2) pieces of 1/2" plate, 8" X 8".

3. (2) pieces of 1/2" X 3" flat bar, 30" long.

4. (1) piece of 1/2" X 3" flat bar, 16" long.

5. (2) pieces of 1/2" sch. 40 black pipe, faced to 3" in length.

6. (1) piece of 2-3/8" diameter CRS, 4" long.

7. (1) piece of 2" sch. 40 black pipe, 13' to 17' long.

8. (1) piece of 1-1/2" sch. 40 black pipe, 13' to 17' long.

9. (1) hollow steel ball, 4" in diameter. Item #13-40-HBL available from King Architectural Metals. Telephone 1-800-542-2379.

10. (4) 1/2" X 2" bolts with nuts and lock washers.

11. (2) 5/8" X 5" bolts with nuts and lock washers.

12. (1) 1/2" X 3" bolt.

13. (1) cast steel boat cleat.

14. Pulley and rope.

A neat bevel or chamfer on a job adds a degree of finished look that is all out of proportion to the effort required to put it there.

in a coastal region, or in a desert region or anywhere where there may be doubts about the ability of the soil to support a flagpole, then check with the local building codes about a foundation. For that reason I only list the materials required to build that part of the flagpole which will be above the ground. And more disclaimers and cautions. It should go without saying that flagpoles and power lines should never, NEVER be in the same place at the same time. If you choose to build and install a flagpole use some common sense in the process.

The materials used in this project were largely leftovers from other jobs but I will say this: I don't think it would be wise to go much beyond 25 or maybe 30 feet in height using the materials I use here. I used 2" schedule 40 black

2. Drilling the locating holes in the 2" pipe.

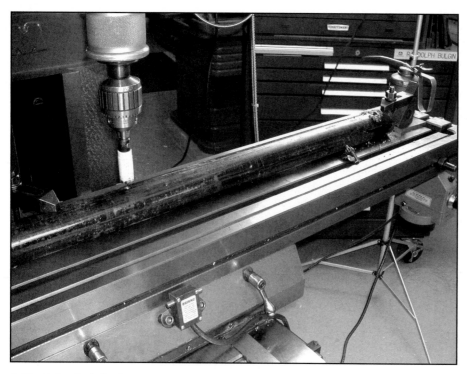

pipe for the bottom 17 feet and 1-1/2" schedule 40 black pipe for the top section. Heavier materials will change the methods of construction and will almost certainly change the way in which it is erected.

3. Machining the holes for the 1/2" pipe.

4. 1/2" pipe sleeve welded into place.

Another thing you will no doubt notice about this chapter is a touch of nostalgia. I started learning the fine art of drafting, or mechanical drawing as we called it in high school, in 1955. At that time the tools were the traditional drawing board, T-square, triangles, protractor, compass and pencils. And lots of erasers! We could get extra credit in class by redrawing our efforts using ink and the bird-beak-looking inking pens of the day, but I usually made a mess of that so I pretty much stuck to pencil. Those rudimentary skills have served me well over the years. I have learned some of the basics of CADAM and of AutoCAD and I have tried some of the less expensive and simpler drawing software packages which abound today, but all of my drawing skills are based on those early days of sitting hunched over a drawing board, Trying to get my assignments completed before the grading period ended. In this chapter I am going back in time to that era. The drawings included here as building aids are all done with the architect's scale, the triangles and pencils. And lots of erasers.

So let's get started. Begin by building what goes into the ground. Apply all of the things I have said about the condition of the soil where you live but you should end up with a piece of 1/2" plate installed at ground level or an inch or so above ground level. See Photo number 1. It should have four 1/2" holes in it, one at each corner. And you should have another piece of 1/2" plate with four holes drilled to match the ones in the plate you have just cemented into the ground. Install the parts which go in the ground at least two days before installing the finished pole so that the concrete used will have cured properly. While that is happening go back to the shop and build the base of the flag pole itself. Using the accompanying photographs as a guide, set the lower section of pipe up in the milling machine. If you are careful you can do this in a drill press but with the dimensional control a milling machine allows it will be easier to do in the mill. Photos 2 and 3 show the process used. Locate the pipe using the slot in the milling machine table as a center. Then drill through in both locations with a 1/4" drill. Don't drill into your milling machine table here or you will never forgive yourself. I won't forgive you either, for that matter! Then, using a hole saw, machine a hole the same size as the outside diameter of a piece of 1/2" pipe through one side only. Do this

5. Ready to weld up the base.

6. Welding the base legs.
7. Welding to the base plate.

8. Assembling the center joint.

for both locations in the same setup.

Without moving the Y-axis of the table, turn the pipe over and line up the hole using the 1/4" pilot drill of the hole saw and machine both holes through the other side of the pipe. You will wind up with two neatly centered holes through the 2" pipe, spaced exactly 24" apart and a good fit up for welding in the 1/2" pipe sleeves. Weld the sleeves in place at this time. See Photo 4.

Now weld up the three pieces of 1/2" X 3" flat bar which make up the lower support for the pole. Photo 5 shows the setup for welding these pieces together. Notice the washers placed in the assembly before welding. These will be removed before installing the pole but they are needed here to provide a clearance between the vertical pieces of the base and the pole itself. Then square the vertical parts to the 8" square base you already have made and finish welding the entire assembly. Refer to Photographs 6 and 7. The two views shown in Drawing number 1 should answer any questions about the construction details.

Now let's move up the center of the pole. There are two options here and they will depend upon whether you are going to install this flag pole within walking distance of your shop or if you have to move it some distance away by truck or trailer. The construction methods are the same. The only difference is whether or not you weld the center joint together OR just slide it together and install the pole. I have made a couple both ways and the single advantage to not welding them together is transportation. Welding the joint will not add any appreciable strength to the construction.

9. Below, the ball and pulley assembly.
10. Right, installing the ball and pulley.

Drawing number 2 shows the assembly of the center section. There are two pieces to machine and about a nickel's worth of steel in both of them. Both pieces weld to the 1-1/2" pipe. Photo 8 show the two sections of the pole being assembled. Part numbers 01 and 02 are the BUSHINGS which will have to be machined from CRS.

11. The erection crew from VFW Post 7339. Left to right is Bill, Bob, Chuck and Phil.

And now on to the top of the pole. I used the hollow steel ball for the top of my flagpole and for others I have built for customers. Many people prefer the eagle as adornment for flagpoles and the supplier I have listed in the materials list is a possible source. I am sure there are others. But whatever you use as a finial will not change the method of construction. You will need to machine two parts for the top, the insert which goes into the top of the 1-1/2" pipe and the sleeve to which is attached the pulley. I had to take my flagpole down at one time to replace the rope and I made a modification which can be seen here in Photo 9. The shielding tab welded to the pulley housing was placed there to keep the rope from getting itself out of the pulley. I don't know how that happens, but it does and this will prevent it. Drawing number 3 and Photo 10 show the attachment of the ball and pulley to the top of the pole. Drawing number 4 is simply details of the parts which require machining.

All that is left is to weld on the boat cleat which you have purchased or possibly made. Weld it on at a convenient height — eye level works pretty good.

Now what are you going to do? You have a 30 to 35 foot long weldment lying there in the middle of your shop with possibly half of it sticking out the door. This is the time when you will appreciate friends. I recruited a flag raising crew from the local VFW Post 7339 to assist me

12. First step towards
raising the pole.

in installing mine and you will need some help to put yours up. Photo 11 is a picture of my crew, but you will have to arrange for your own.

I am not going to describe in great detail how to install the completed flagpole, but there are a few things I will mention which will make it easier. After you have completed construction of all of the components and painted them with as good an outdoor paint as is available at your local hardware store you will begin by bolting the base assembly to the plate you have already put in the ground. Then place the top sleeve in position as seen in Photo 12 and put the bolt in but do not tighten the nut yet. Next — yada, yada, yada — get the pole upright and put the bolt in the bottom sleeve and tighten both bolts.

Be careful! The completed pole, if you made it to these dimensions will weigh in the neighborhood of 120 to 130 pounds. That is not a lot of weight but imagine trying to haul in a fish weighing 130 pounds on a 35 foot long fishing pole. You will definitely need help! And the warning at the beginning of the chapter bears repeating. Be aware of power lines!

I try to put the flag up every morning and take it down every evening. I usually don't fly it in really bad weather and I sometimes forget to take it down at night, but I get a great deal of satisfaction from seeing the flag displayed in front of my shop. I defended it while in the U.S. Navy as a young man and I will defend it now as a "mature" metal worker. Don't mess with my flag!

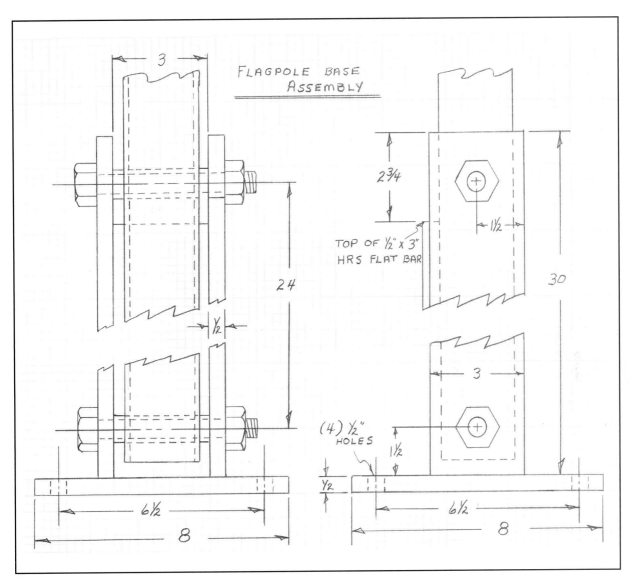

FLAGPOLE BASE ASSEMBLY

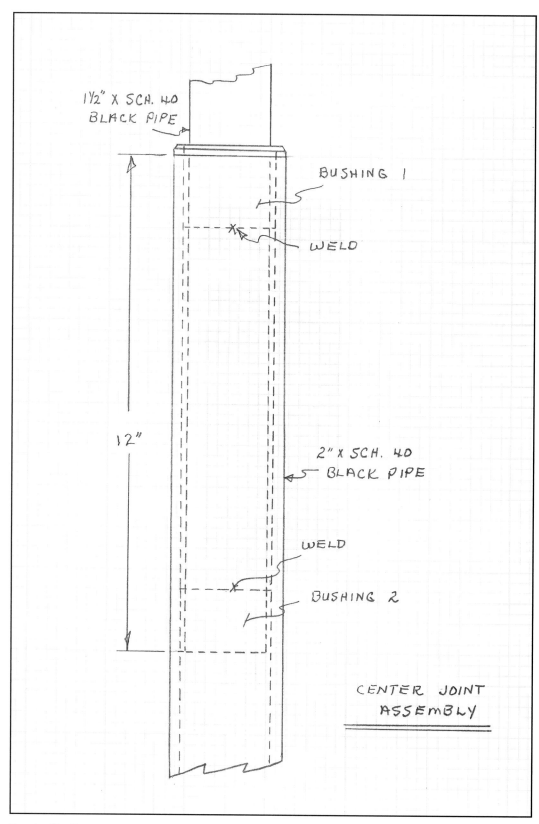

Drawing No. 2.

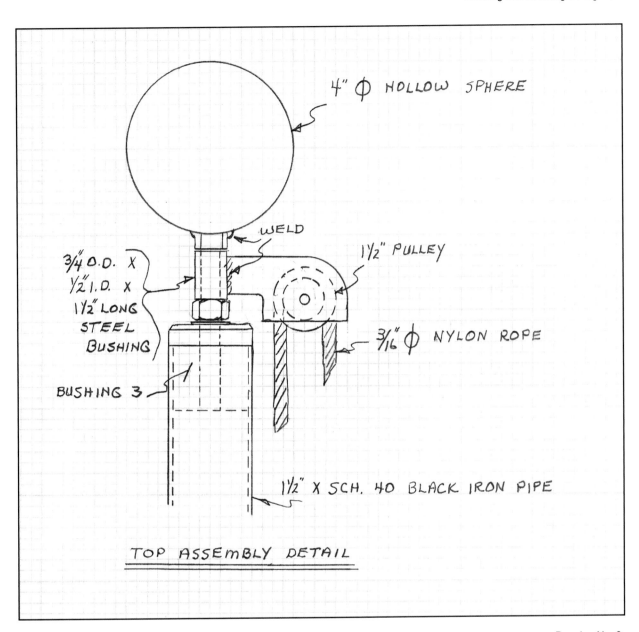

4" ⌀ HOLLOW SPHERE

WELD

1½" PULLEY

¾" O.D. X ½" I.D. X 1½" LONG STEEL BUSHING

3/16" ⌀ NYLON ROPE

BUSHING 3

1½" X SCH. 40 BLACK IRON PIPE

TOP ASSEMBLY DETAIL

Drawing No. 3.

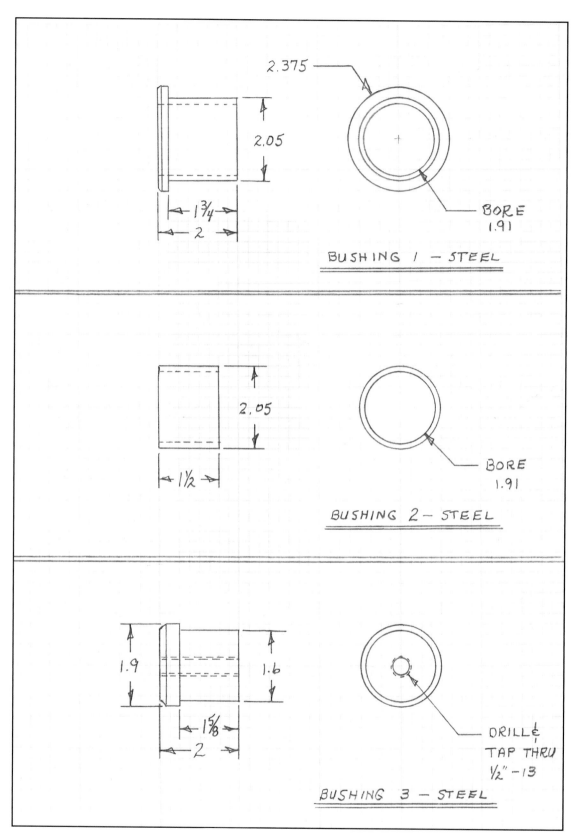

Drarwing No. 4.

Chapter 15

Making Eggs

Every now and then I get the urge to go to the shop and make something that, when it is completed, has absolutely no value. Something that I just make because it is fun to make. I will agree that making something just because it is fun does have the value of being fun. And these days that is not to be taken lightly. Probably one of the most important, if not the most important, benefits of a hobby is that it allows you to spend time and not have to justify the expenditure. You do whatever it is that you do for the pure pleasure of doing it. There isn't the pressure of having to demonstrate usefulness or profitability or necessity as justification. You just like to do it. As an added value this project also brought some pleasure to my shop's banana pudding making department and that never hurts. So let's make some eggs! Or lay some eggs, as the case may be.

Making one egg "just alike" is no problem. The difficulty in a project like this is making several oval profiles all just alike without the benefit of a CNC lathe or machining center. The process used here has the added advantage of being a useful system for producing other profiles which can be turned on a lathe, but which do not

1. Machining the material to 1" diameter. The material in this photo is cast iron.

2. Machining the profile of the oval end of the egg.

have clearly defined dimensional properties in terms of lengths, angles and diameters. The making of machine handles is one such example and I am sure there are others. There is an acquired talent for freehand profiling, both on the lathe and on the milling machine, but it is something that, without practice, can be frustrating. You might even have a need for this process in making something far more useful than the project described here. Whether you ever have use for this method of reproducing shapes or not it is a good thing to know. And the fact remains that making these eggs is fun.

We will begin by chucking up a piece of material and turning it to an appropriate length and diameter as shown in Photo 1. An egg has a pleasing profile in that it has a proportionate length to diameter ratio. A ratio of 1.5:1 works well here. In this case I have chosen to make the diameter of the eggs 1" with an overall length of 1.5."

Turn the 1" diameter back to a length of approximately 2." This will give you room to machine the end nearest to the chuck without crowding yourself when you introduce the file to that end. Starting at a point 1" towards the chuck from the end of the part, machine the oval end of the egg to a profile which is pleasing to you. It is difficult to describe exactly what you are looking for here, but you will know it when you get it. This will require some hand work and the use of a long angle file to get the exact results you are looking for. See Photo 2. Remove all of the tool marks from the part but do not polish the profile yet.

Now cover the part with layout dye and we are ready to make our chart for making other eggs "eggzactly" like this one. Position a tool with a sharp point. I used the corner of the parting tool, but any tool which gives you a definite point will work, so that the marking point is exactly 1.000" from the end of the part and move the tool in until you just make a visible mark on the profile. Set your DRO to zero/zero at this point. If you do not have a DRO then position your dial indicators in both X and Z axes so that they read zero/zero. This is your starting point. Move the carriage .025" towards the tail stock and then move the cross slide in until you make a mark on the part. Record the resulting readings as your second point. Then move another .025" towards the tailstock and repeat the process. Continue with this as shown in Photo 3 until you have moved the full 1" length to the end of the part.

If the profile of the egg shown in Photo 4 is pleasing to you and if you do not want to take the time to develop your own chart, use the chart at right. The round end of the egg is a true radius and will

Egg profile	
Z	X
.000	.000
.025	.001
.050	.002
.075	.004
.100	.006
.125	.009
.150	.013
.175	.016
.200	.019
.225	.023
.250	.028
.275	.032
.300	.037
.325	.042
.350	.047
.375	.052
.400	.058
.425	.064
.450	.072
.475	.079
.500	.087
.525	.095
.550	.103
.575	.112
.600	.121
.625	.130
.650	.148
.675	.156
.700	.164
.725	.176
.750	.188
.775	.202
.800	.215
.825	.235
.850	.252
.875	.273
.900	.295
.925	.322
.950	.356
.975	.409
1.000	.500

3. Finding the co-ordinates for the oval end.

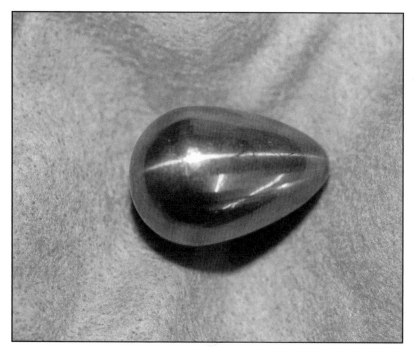

4. Copper egg showing the profile.

5. Machining the round end of the egg.

be generated by using the wonderful little booklet by Guy Lautard entitled *Tables and Instructions for Ball and Radius Generation.*

You can make your own radius chart using either the process described for the oval end of the egg or by developing the co-ordinates using trig formulas, but this booklet will save you a lot of time and it includes tables for generating radii from 1/64" up through 2." Photo 5 shows the radius almost completed. Be aware of your material here. If you are working with cast iron or with some of the more fragile synthetic materials you will not want to reduce the diameter to a point where it could break off before you are finished with it. You should be finished with the oval end of the egg, including polishing, before you go too far with the round end.

Begin the polishing process by first making sure all of the gross tool marks have been removed with a file. Then polish with abrasive cloth starting with about 180-grit and going through progressively finer grits until you are satisfied with the result. When you are using abrasive cloth or paper be sure that you remove all of the surface marks from the preceding grit before you go to the next. The clear acrylic egg in Photo 6 was finished with a 1500-grit wet or dry sanding paper available from hardware stores or from automotive body repair shops. Keep the paper wetted with plain water for the final finishing process. Photo 7 shows the cast iron egg polished on the pointed end and ready to be parted off.

Make a work holding fixture for polishing the radiused end of the egg by drilling into a piece of nylon or other material which

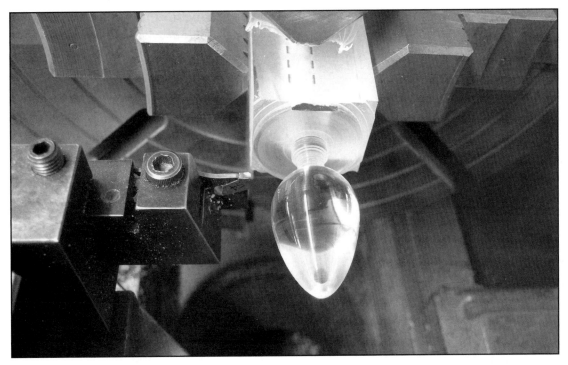

6. A see-through egg of cast acrylic.

will not scratch or mar the polished surface. Drill a hole 1/2" in diameter to a depth of a little over 1" and then drill with a 1" bit to a depth of approximately 0.7." The 159-degree drill point will produce a pocket which will automatically center the part when it is pushed all the way into the hole. Cut three notches around the periphery of the fixture with a hand hacksaw and put it into the three jaw chuck. Photos 8 and 9 show the fixture being used to finish the round end of the cast iron egg while Photo 10 is of the fixture by itself.

It has been entertaining to find how many different materials can be used to make eggs and how many different colors and textures you will have in your egg basket as a result. The basket shown on the back cover has eggs made from the following materials:

7. Polishing the oval end of the cast iron egg.

1. Metals: 6061 Aluminum, 360 brass, 630 bronze, G2 gray cast iron, 145 copper, O1 tool steel, and 303 stainless steel. Nice colors may be obtained from the O1, or from CRS for that matter, by putting it in the furnace or in a bed of hot sand at temperatures ranging from a little over 400 degrees F up to 600 or 700 degrees F. The colors will range from a pale yellow at the lower temperatures to a beautiful blue tint at the upper ranges.

2. Wood: American

black walnut, maple, Lignum Vitae.

3. Synthetic materials: Clear Cast Acrylic, Black Polycarbonate, 6/6 White Nylon, Blue Nylon, PEEK (Polyetheretherketone), Polyimide, Garolite and Teflon. There is also an egg made from a piece of amber colored material a friend of mine brought me which is some sort of a synthetic resin, but I have no idea of its composition.

And here is a good place to put the usual caveat. When you are working with synthetic materials, and with some natural materials, be aware that there can be some noxious and possibly dangerous fumes generated from the process. As is always the rule — know your material. Work in a well ventilated area. Wear the appropriate protective filtering mask. I hope we don't have to mention safety glasses again. If you have doubts about any material you are working with go to the trouble of finding and reading the MSDS for that material. This of course applies to whether you are doing something as useless as making eggs or as useful as repairing a truck axle.

I have often thought that it would make an interesting project to see just how many colors and textures it would be possible to obtain in a basket of eggs like this. *Machinery's Handbook* has a whole section on the coloring of metals using various processes. Plating, anodizing, Parkerizing, bluing, etching and many other processes could be included as well as the many different chemical

8. Top, polishing the round end.
9. Above, polishing completed.

changes which can be effected. Color case hardening also has potential. One of these days when I have absolutely nothing else to do I may set in to do it. After I get the fourteen major projects and the 350 or so minor jobs which I either have already started or will start on any day now, of course.

10. Nylon polishing fixture.

It is hard for me to make this project sound anything at all like a useful way to pass the time, but if there is a redeeming characteristic to the job it is this. This nest of eggs has caused probably more comments and questions from visitors to my shop than any other thing I have lying around and I always thought I had some pretty interesting stuff in my shop. There will be some who see this and think that this guy just has too much time on his hands. But it was fun to do and it is fun to talk about and having a little fun is a really important thing these days.

APPENDIX

Acronyms and Abbreviations

CNC	Computer Numerical Control
CRS	Cold Rolled Steel
DCEN	Direct current electrode negative
DCEP	Direct current electrode positive
FHCS	Flat head cap screw
GMAW	Gas metallic arc welding
GTAW	Gas tungsten arc welding
HAZ	Heat affected zone
HRS	Hot Rolled Steel
HSS	High Speed Steel
MIG	Metallic inert gas (Same as GMAW)
MSDS	Material Safety Data Sheet
O/A	Oxygen Acetylene
SAW	Submerged arc welding
SMAW	Shielded metallic arc welding (Plain old stick welding)
SPC	Statistical Process Control
TIG	Tungsten inert gas (Same as GTAW)
TPG	Turned, ground and polished steel shafting
SHCS	Socket head cap screw